STUDENT SOLUTIONS MANUAL FOR

OTT/LONGNECKER'S

AN INTRODUCTION TO STATISTICAL METHODS AND DATA ANALYSIS

Fifth Edition

Michael Longnecker
Texas A & M University

DUXBURY

™

THOMSON LEARNING

Australia • Canada • Mexico • Singapore • Spain • United Kingdom • United States

Sponsoring Editor: *Carolyn Crockett*
Assistant Editor: *Ann Day*
Editorial Assistant: *Jennifer Jenkins*
Marketing Manager: *Joe Rogove*
Marketing Associate: *Maria Salinas*
Production Editor: *Scott Brearton*
Cover Design: *Laurie Albrecht*
Print Buyer: *Chris Burnham*
Cover Printing: *Webcom Limited*
Printing and Binding: *Webcom Limited*

For more information about this or any other Duxbury products, contact:
DUXBURY
511 Forest Lodge Road
Pacific Grove, CA 93950 USA
www.duxbury.com
1-800-423-0563 (Thomson Learning Academic Resource Center)

For permission to use material from this work, contact us by
www.thomsonrights.com
fax: 1-800-730-2215
phone: 1-800-730-2214

Printed in Canada
10 9 8 7 6 5 4 3 2 1

ISBN: 0-534-37123-X

TABLE OF CONTENTS

Chapter 1-What is Statistics   1

Chapter 2-Using Surveys and Scientific Studies to Gather Data   2

Chapter 3-Data Description   4

Chapter 4-Probability and Probability Distributions   19

Chapter 5-Inferences about Population Central Values   31

Chapter 6-Inferences Comparing Two Population Central Values   44

Chapter 7-Inferences about Population Variances   59

Chapter 8-Inferences about More Than Two Population Central Values   70

Chapter 9-Multiple Comparisons   75

Chapter 10-Categorical Data   78

Chapter 11-Linear Regression and Correlation   89

Chapter 12-Multiple Regression and the General Linear Model   110

Chapter 13-More on Multiple Regression   129

Chapter 14-Design Concepts for Experiments and Studies   155

Chapter 15-Analysis of Variance for Standard Designs   157

Chapter 16-Analysis of Covariance   173

Chapter 17-Analysis of Variance for Some   179
             Fixed-, Random-, and Mixed-Effects Models

Chapter 18-Repeated Measures and Crossover Designs   187

Chapter 19-Analysis of Variance for Some Unbalanced Designs   194

Chapter 1

What Is Statistics

1.1 a. The population of interest is the weight of shrimp maintained on the specific diet for a period of 6 months.

     b. The sample is the 100 shrimp selected from the pond and maintained on the specific diet for a period of 6 months.

     c. The weight gain of the shrimp over 6 months.

     d. Since the sample is only a small proportion of the whole population, it is necessary to evaluate what the mean weight may be for any other randomly selected 100 shrimps.

1.5 a. All football helmets produced by the five companies over a given period of time.

     b. The 540 helmets selected from the output of the five companies.

     c. The amount of shock transmitted to the neck when the helmet's face mask is twisted.

     d. The neck strength of players is extremely variable for high school players. Hence, the amount of damage to the neck varies considerably from player to player for exactly the same amount of shock transmitted by the helmet.

Chapter 2

Using Surveys and Scientific Studies to Gather Data

2.1 The relative merits of the different types of sampling units depends on the availability of a sampling frame for individuals, the desired precision of the estimates from the sample to the population, and the budgetary and time constraints of the project.

2.3 A more precise estimate can be obtained by considering individual cars but it may be very difficult obtaining the sampling frame. By selecting parking lots and examining all cars in the lot, the data is more easily obtained but the individual cars in the lot may have common characteristics reflecting the set of persons using the parking lot. Thus, the cars in the lot are a cluster sample and not a simple random sample. This results in a less precise estimate of the population than examining the same number of cars selected individually.

2.5 The agency could stratified farms based on the total acreage of farms in the state. A simple random sample of farms could then be selected within each strata and a questionaire sent to the farmer.

2.9   a. No. The survey in which the interviewer showed the peanut butter should be the more accurate because it does not rely on the respondent's memory of which brand was purchased.

     b. Both surveys may have survey nonresponse bias because an entire segment of the population (those not at home) cannot be contacted. Also, both surveys may have interviewer bias resulting from the way the question is posed (e.g., tone of voice). In the first survey, results may be biased by the respondent's ability to recall correctly which brand was purchased. The second survey may be biased by the respondent's unwillingness to show the interviewer the peanut butter jar (too intrusive), or by the respondent not recognizing that the peanut butter that had purchased was *low fat*.

2.11   a. "Employee" should refer to anyone who is eligible for *sick days*.

     b. Use payroll records. Stratify by employee categories (full-time, part-time, etc.), employment location (plant, city, etc.), or other relevant subgroup categories. Consider systematic selection within categories.

     c. Sex (women more likely to be care givers), age (younger workers less likely to have elderly relatives), whether or not they care for elderly relatives now or anticipate doing in the near future, how many hours of care they (would) provide (to define "substantial"), etc. The company might want to explore alternative work arrangements, such as flex-time, offering employees 4 ten-hour days, cutting back to $\frac{3}{4}$-time to allow more time to care for relatives, etc., or other options that might be mutually beneficial and provide alternatives to taking sick days.

2.13   If phosphorus first: [P,N]

[10,40], [10,50], [10,60], then [20,60], [30,60]

Or [20,40], [20,50], [20,60], then [10,60], [30,60]

Or [30,40], [30,50], [30,60], then [10,60], [20,60]

If nitrogen first: [N,P]

[40,10], [40,20], [40,30], then [50,30], [60,30]

Or [50,10], [50,20], [50,30], then [40,30], [60,30]

Or [60,10], [60,20], [60,30], then [40,30], [50,30]

Chapter 3: Data Description

3.3 Graphical Methods

3.1   a. Pie Chart

   b. Bar Graph

3.3 Pie Chart and Bar Graph should be plotted. Pie chart is better since we are proportionally allocating the food dollar to seven categories.

3.7 Two separate bar graphs could be plotted, one with Lap Belt Only and the other with Lap and Shoulder Belt. A single bar graph with the Lap Belt Only value plotted next to the Lap and Shoulder for each value of Percentage of Use is probably the most effective plot. This plot would clearly demonstrate that the increase in number of lives saved by using a shoulder belt increased considerably as the percentage use increased.

3.9 The relative frequency histogram is multimodal, skewed to the right.

3.11 The stem-and-leaf would be since the exact values for the 24 cities is being displayed, whereas in the relative frequency histogram the data is grouped into classes.

Stem-and-Leaf Plot	(Leaf Unit 100)
2	70,71,98
3	59,60,82
4	51,67,88,98
5	12,20,61,72
6	34,43,86
7	84
9	47,86
13	05
16	82
24	70
49	04

The stem-and-leaf plot is more informative because it provides the exact values of the data as well as the frequencies within each stem.

3.13   a. Construct separate relative frequency histograms.

   b. The histogram for the New Therapy has one more class than the Standard Therapy. This would indicate that the New Therapy generates a few more large values than the Standard Therapy. However, there is not convincing evidence that the New Therapy generates a longer survival time.

3.15   a. The outlays have increased rapidly from 1980 to 1989, then dropped somewhat until 1991, when they increased again to 1992. The values had a slight decrease from 1992 through 1997.

b. The % GNP increased rapidly form 1980 to 1985, remain steady for a couple of years, then decreased fairly rapidly from 1987 through 1997 (except a minor jump from 1991 to 1992).

c. The two plots have very different trends. The public interest group's contention is not supported by either graph for the decade of the 90's.

3.19 There are a few states having either a very small or very large number of telephones per thousand residents. The vast majority of states have values in the 450 to 700 range.

3.23 The plots show an upward trend from year 1 to year 5. There is a strong seasonal (cyclic) effect; the number of units sold increases dramatically in the late summer and fall months.

3.4 Measures of Central Tendency

3.25 Mean $= 243/16 = 15.1875 = 15.2$, $\qquad$ Median $= (14+15)/2 = 14.5$, $\qquad$ Mode $= 18$

3.27 10% of measurements $= 10\%$ of $16 = 1.6 \approx 2$. Thus, we delete the two largest and smallest data values. Since the 2 largest values are deleted in computing the 10% Trimmed Mean, its value is the same for both data sets:

10% Trimmed Mean $= (11+11+12+13+14+14+15+17+17+18+18+18)/12 = 14.83$. Thus, the extreme values do not have an effect. A 5% Trimmed Mean deletes only the smallest and largest values in the data set. Therefore, the second largest extreme value does enter into the calculation and hence would have an effect on the 5% Trimmed Mean. The effect would not be as dramatic as was seen in using the untrimmed mean.

3.29 The following table is used to calculate the summary statistics:

Class Interval	Frequency (f_i)	Midpoint(y_i)	$f_i y_i$
0-2	1	1	1
3-5	3	4	12
6-8	5	7	35
9-11	4	10	40
12-14	2	13	26
Total	15		114

mean $\approx 114/15 = 7.6 \qquad$ mode ≈ 7

median $\approx L + \frac{w}{f_m}(.5n - cf_b) = 6 + \frac{2}{5}[(.5)(15) - 4] = 7.4$

3.31 a. The mean cannot be approximated since we do not know the endpoint for the last interval, hence cannot compute the midpoint of this interval. Mode ≈ 240.

median $\approx L + \frac{w}{f_m}(.5n - cf_b) = 200 + \frac{19.9}{88}[(.5)(408) - 119] = 219.2$

b. The median since it would give an indication of the cholestrol level for which half of men in this group have a greater of lesser value.

5

3.33 a. After removing the survival times of the two individuals who left the study, we obtain Mean = 46.3. The median can be calculated for all 11 patients, since we know that the values for the two indiduals who left the study were greater than the listed values of 57 and 60 which would place them in the upper half of the survival times. Thus, we obtain Median = 29.

b. The median would be unchanged but the mean would increase since these two values will be greater than the mean calculated from the nine observed values.

3.35 a. The rounded measurements:

2.10	2.80	2.17	1.99	2.22	3.09
2.40	2.50	2.80	2.10	2.92	2.20
1.70	2.70	2.82	2.67	3.05	2.93
2.90	1.90	2.38	2.65	2.77	1.85
1.60	2.70	2.68	2.06	2.36	2.28
2.70	2.40	2.39	2.55	1.80	1.96

b. Sample modes = 2.10, 2.77, 2.82

c. Median = $(2.43 + 2.47)/2 = 2.45$

d. Mean = 2.43

3.37 a. The values are given below:

Group	Mean	Median	Mode
I	2.923	2.805	no mode
II	1.592	1.565	1.55, 1.57
III	0.797	0.755	0.70

b. mean = 1.7707 median = 1.565 mode = 0.70, 1.55, 1.57

c. If we were to use one summary for the combined group, then the median would be most appropriate because the three groups are substantially different. If separate summaries are computed for each group, then the mean and median are both appropriate since the three groups have relatively symmetric distributions.

3.5 Measures of Variability

3.39 a. $\bar{X} = \sum y_i/n = \frac{40}{8} = 5$. Yes.

b. $\sum_{i=1}^{8}(y_i - 5)^2 =$
$(6 - 5)^2 + (3 - 5)^2 + (10 - 5)^2 + (4 - 5)^2 + (4 - 5)^2 + (2 - 5)^2 + (4 - 5)^2 + (7 - 5)^2 =$
$1 + 4 + 25 + 1 + 1 + 9 + 1 + 4 = 46$

c. $s^2 = \frac{46}{8-1} = 6.57 \Rightarrow s = 2.56$. The $CV = 100\frac{s}{\bar{y}}\% = 100\frac{2.56}{5}\% = 51\%$. The standard deviation is 51% of the mean.

3.41 Age: $s \approx (10 - 2)/4 = 2$. The estimate is fairly close to the actual value of 2.56.

Experience: $s \approx (39 - 18)/4 = 5.25$. The estimate is somewhat smaller than the actual value of 7.95.

3.43 The quantile plot is given here.

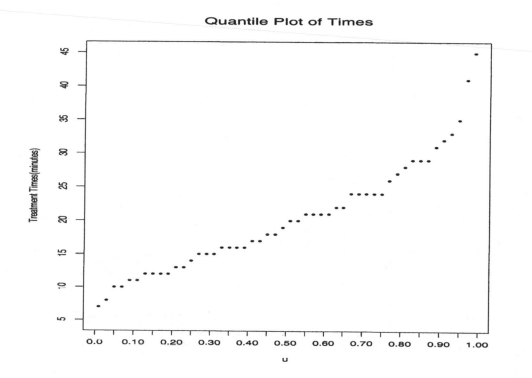

a. The 25th percentile is the value associated with $u = .25$ on the graph which is 14 minutes. Also, by definition 14 minutes is the 25th percentile since 25% of the times are less than or equal to 14 and 75% of the times are greater than or equal to 14 minutes.

b. Yes; the 90th percentile is 31.5 minutes. This means that 90% of the patients have a treatment time less than or equal to 31.5 minutes (which is less than 40 minutes).

3.45   a. Range ≈ 200

b. $s \approx 200/4 = 50$ or we can use the method of estimating s from grouped data which yields

$s^2 \approx \frac{1}{190}((10 - 96.2)^2 6 + (30 - 96.2)^2 11 + (50 - 96.2)^2 16 + (80 - 96.2)^2 59 +$

$+(110 - 96.2)46 + (130 - 96.2)^2 33 + (150 - 96.2)^2 16 + (180 - 96.2)^2 4)) = 1385.716$

$s \approx \sqrt{1385.716} = 37.2$

c. $\bar{y} \pm s$ yields $(59, 133.4)$; $121/191 \approx 64\%$

$\bar{y} \pm 2s$ yields $(21.8, 170.6)$; $181/191 \approx 94.8\%$

$\bar{y} \pm 3s$ yields $(-15.4, 207.8); 191/191 = 100\%$

The empirical rule works well for this data set since the relative frequency histogram is mound-shaped.

3.47   a. The time series plot is given here.

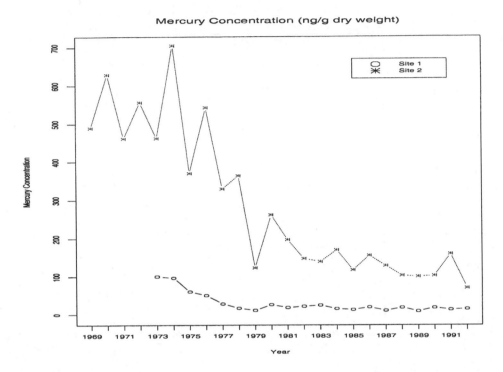

For Site 1, there is a steady decrease until 1980, after which the level is fairly constant but at a much lower level than the values for Site 2. The concentrations at Site 2 are very erratic from 1969 to 1980, with alternating rises and falls. From 1980 through 1992, there is a fairly steady decline in mercury concentration.

b. Site 1: Median = 18.25,   Mean = 29.18

Site 2: Median = 184.1,   Mean = 287.1

Both distributions are right skewed, thus the median is a more appropriate measure of center than is the mean. Site 1 has a considerably lower center than that of Site 2.

c. Site 1: s = 26.95, CV = 92%

Site 2: s = 194.7, CV = 68%

Comparing the standard deviations can be misleading because 26.95 is smaller in magnitude than 194.7. However, the data values for Site 1 are considerably smaller in magnitude also. Therefore, it is more informative to compare the CV values. Based on CV values the concentrations from Site 1 are relatively more variable than those from Site 2.

d. No, Site 1 does not have values for these years.

3.6 The Boxplot

3.49 $Q_2 = 20$ $Q_1 = (17+18)/2 = 17.5,$ $Q_3 = (23+24)/2 = 23.5$

3.50   a. Stem-and-Leaf Plot is given here:

Stem	Leaf
2	5
2	677
2	9
3	00111
3	223333
3	5
3	67
3	8

b. Min=250, $Q_1 = (295 + 301)/2 = 298, Q_2 = (315 + 320)/2 = 317.5,$
$Q_3 = (334 + 334)/2 = 334,$ Max=386

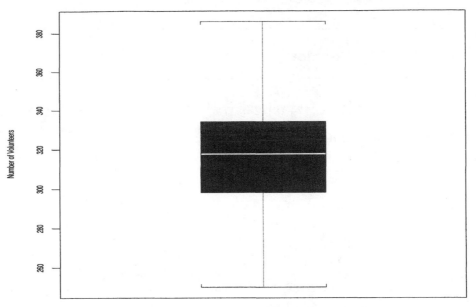

Number of Blood Donors On Fridays

There are no outliers because $Q_1 - (1.5)IQR = 244 < 250$ and $Q_3 + (1.5)IQR = 388 > 386$. The distribution is approximately symmetric with no outliers.

3.51   a. CAN: $Q_1 \approx 1.45$, $Q_2 \approx 1.65$, $Q_3 \approx 2.4$
       DRY: $Q_1 \approx 0.55$, $Q_2 \approx 0.60$, $Q_3 \approx 0.7$

   b. Canned dogfood is more expensive (median much greater than that for dry dogfood), highly skewed to the right with a few large outliers. Dry dogfood is slightly left skewed with a considerably less degree of variability than canned dogfood.

3.7 More Than One Variable

3.52   a. Stacked Bar Graph is given here:

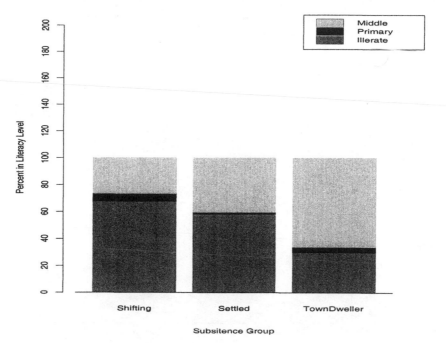

Literacy Level of Three Subsistence Groups

b. Illiterate: 46%, Primary Schooling: 4%, At Least Middle School: 50%

Shifting Cultivators: 27%, Settled Agriculturists: 21%, Town Dwellers: 51%

There is a marked difference in the distribution in the three literacy levels for the three subsistence groups. Town dwellers and shifting cultivators have the reverse trends in the three categories, whereas settled agriculturists fall into essentially two classes.

3.53 Percentage comparison based on Row Totals:

Turnover	Age(years)				
Reason	< 29	30-39	40-49	≥ 50	Total
Resigned	50	10	6.7	33.3	100(n=60)
Transferred	18.2	68.2	6	7.6	100(n=66)
Retired/fired	6.45	7.26	41.94	44.35	100(n=124)

50% of workers who resign are in the youngest age group; of those who transfer, 68.2% are in their 30's; and of those who either retire or got fired 86.24% are at least 40 years old.

3.55   a. The means and standard deviations are given here:

Supplier	$\bar{y}$	s
1	189.23	2.96
2	156.28	3.30
3	203.94	8.96

11

b. Side-by-side Boxplots are given here:

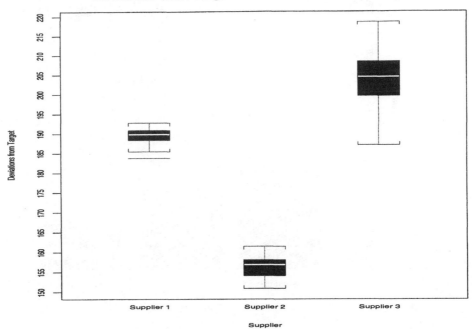

The three distributions are relatively symmetric but supplier 3 is considerably more variable and is shifted above supplier 1's values, which in turn are shifted above supplier 2's values.

c. Supplier 3 not only has the largest mean but also the largest standard deviation. Suppliers 1 and 2 have similar degrees of variability but supplier 1 has a greater mean than supplier 2.

d. Supplier 2 because it has the smallest mean and deviates with essentially the same degree of variability as supplier 1.

3.57 A time series plot with M2 and M3 values on the vertical axis and months on the horizontal axis is given here.

12

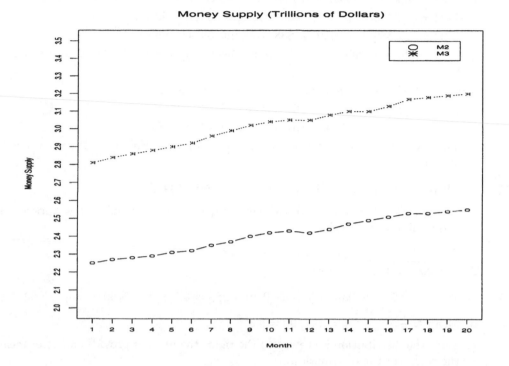

Money Supply (Trillions of Dollars)

This is a more informative plot than the scatterplot because it shows the relative changes of the two measures of money supply across the 20 months.

Supplementary Exercises

3.59　a. mode = 5,　　　median = 15,　　　mean = 15.96

　　　b. range = 34-4 = 30,　　　$s \approx 30/4 = 7.5$

　　　c. s = 8.5

　　　d. No, the histogram for the data set is skewed to the right and hence is not mound-shaped.

3.63　a. Mean = 0.65;　　　Median = 0.55;　　　Mode = 0.5

　　　b. Mean = 1.21;　　　Median = 0.55;　　　Mode = 0.5

　　　The mean increases but the median and mode remain the same.

3.65　a. Construct a relative frequency histogram

13

b. Highly skewed to the right

c. $1424 \pm 3488 \Rightarrow (-2063, 4912)$ contains $37/41 = 90.2\%$

$1424 \pm (2)3488 \Rightarrow (-5551, 8400)$ contains $38/41 = 92.7\%$

$1424 \pm (3)3488 \Rightarrow (-9039, 11888)$ contains $39/41 = 95.1\%$

These values do not match the percentages from the Empirical Rule: 68%, 95%, and 99.7%.

d. $1.48 \pm 1.54) \Rightarrow (-0.06, 3.02)$ contains $31/41 = 75.6\%$

$1.48 \pm (2)1.54 \Rightarrow (-1.60, 4.56)$ contains $40/41 = 97.6\%$

$1.48 \pm (3)1.54 \Rightarrow (-3.14, 6.10)$ contains $41/41 = 100\%$

These values closely match the percentages from the Empirical Rule: 68%, 95%, and 99.7%.

3.67   a. There has been very little change from 1985 to 1996.

b. Yes; For both 1985 and 1996, DC had extremely low ownership. New York and Hawaii had semi-low ownership.

c. No

d. The cost of homes is very high.

3.69   a. Mode = 2.5   Median $\approx L + \frac{w}{f_m}[0.5n - cf_b] = 5.5 + \frac{2}{13}[0.5(90) - 35] = 7.04$

b. Mean $\approx \frac{1}{n}\sum_{i=1}^{13} f_i y_i = 747/90 = 8.3$

c. Since the distribution is skewed to the right, the median provides a better measure of the center of the distribution.

3.71 The quantile plots are given here.

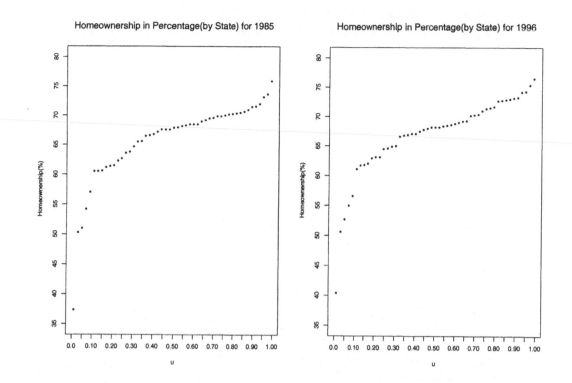

Homeownership in Percentage(by State) for 1985   Homeownership in Percentage(by State) for 1996

a. The 20th percentile for 1996 is given by reading the vertical value on the graph for $u = 0.20 \Rightarrow 63\%$. Thus, approximately 20% of the 1996 homeownership percentages are less than or equal to 63%.

b. The upper 10th percentile would correspond to the states having the largest 5 percentages: Michigan, Indiana, West Virginia, Minnesota, and Maine.

c. In 1985, the states falling in the upper 10th percentile are Pennsylvania, South Carolina, Wyoming, Maine, and West Virginia. There are only two states which fall into both groups.

3.73   a. mode $= 1.0$ median $= 1.0$

b. mean $= \sum_{i=0}^{10} f_i y_i = 663/500 = 1.326$

c. Because the mean is larger than the mode and median, the distribution is skewed to the right.

3.75   a. Relative frequency histogram

b. Mean $\approx \frac{1}{n} \sum_{i=1}^{11} f_i y_i = \frac{1}{50}[(1)(52) + (4)(67) + \cdots + (1)(202)] = 5480/50 = 109.6$

$s^2 \approx \frac{1}{n-1} \sum_{i=1}^{11} f_i (y_i - 109.6)^2 = \frac{1}{49}[47412] = 967.592$

$s \approx \sqrt{967.592} = 31.106$

3.77   a. Plot relative frequency histogram.

b. 1100

c. mean $= 25115/23 = 1091.96,$ median $= 1039$

d. Because the mean is slightly larger than the median, it is likely that the distribution is slightly skewed to the right.

3.79 The stem-and-leaf diagram is given here:

Stem	Leaf
54	7
55	
56	
57	
58	
59	
60	
61	
62	5
63	0
64	
65	6
66	4 7 7
67	79
68	8 88 88
69	1 4 7 9
70	0 1 2 333 8
71	1

Yes, the distribution is left skewed.

3.81 The means and medians are given here:

Number of Members	Mean	Median
1	93.75	78.5
2	98.652	95
3	113.3125	100
4	124.857	112.5
5+	131.90	128.50

3.83 a. New policy $\bar{y} = 2.27,$ $s = 3.26$
Old policy: $\bar{y} = 4.60,$ $s = 2.61$

b. Both the average number of sick days and the variation in number of sick days have decreased with new policy.

3.85 a. The sample mean will be distorted by several large values which skew the distribution. State 5 and State 11 have more than 10 times as many plants destroyed as any other state; for arrests, States 1, 2, 8 and 12 exceed the other arrest figures substantially.

b. Plants: $\bar{y} = 10166919/15 = 677794.60$
Arrests: $\bar{y} = 1425/15 = 95$
10% trimmed mean:

16

Plants: $\bar{y} = 1565604/11 = 142327.64$

Arrests: $\bar{y} = 657/11 = 59.7$

20% trimmed mean:

Plants: $\bar{y} = 1197354/9 = 133,039.33$

Arrests: $\bar{y} = 372/9 = 41.30$

For plants, the 10% trimmed mean works well since it eliminates the effect of States 5 and 11. For arrests, the means differ because each takes some of the high values out of calculation. It appears that the distribution is not skewed, but rather separated into at least two parts: states with high number of arrests and states with low numbers. By trimming the mean, we may be losing important information.

3.87   a. A time series plot is given here:

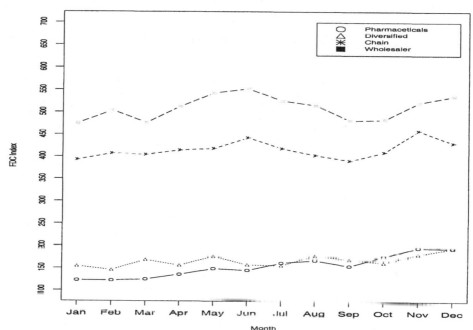

b. All the components vary throughout the year with a slight increasing trend. Wholesalers show the greatest degree of variability through the year.

3.89   a. Average Price — 76.68

     b. Range = 202.69

     c. DJIA = 10856.2

     d. Yes; The stocks covered on the NYSE.

       No; The companies selected are the leading companies in the different sectors of business.

3.91 Only the District of Columbia with 37.4% and 40.4% homeownership in 1985 and 1996, respectively, fell more than 3 standard deviations from the mean. The stem-and-leaf plot showed this value at the low end of the scale, but did not really separate it too much from the rest of the data. It is not too extreme and should not decrease the mean too much. The raw and 10% trimmed means are given here:

Year	Raw Mean	10% Trimmed Mean
1985	65.88	66.91
1996	66.84	67.81

3.93   a. The value 62 reflects the number of respondents in coal producing states who preferred a national energy policy that encouraged coal production. The value 32.8 is of those who favored a coal policy, 32.80% came from major coal producing states. The value 41.3 tells us that 41.3% of those from coal states favored a coal policy. And the value 7.8 tells us that 7.8% of all the responses come from residents of major coal producing states who were in favor of a national energy policy that encouraged coal production.

  b. The column percentages because they displayed the distribution of opinions within each of the three types of states.

  c. Yes. For both the coal and oil-gas states, the largest percentage of responses favored the type of energy produced in their own state.

Chapter 4: Probability and Probability Distributions

4.1 How Probability Can Be Used in Making Inferences

4.1   a. Subjective probability

       b. Relative frequency

       c. Classical

       d. Relative frequency

       e. Subjective probability

       f. Subjective probability

       g. Classical

4.2 Finding the Probability of an Event

4.3 Using the binomial formula, the probability of guessing correctly 15 or more of the 20 questions is 0.021.

4.4 Conditional Probability and Independence

4.5   HHH, HHT, HTH, THH, TTH, THT, HTT, TTT

4.7   a. $P\left(\overline{A}\right) = 1 - \frac{3}{8} = \frac{5}{8}$

         $P\left(\overline{B}\right) = 1 - \frac{7}{8} = \frac{1}{8}$

         $P\left(\overline{C}\right) = 1 - \frac{1}{8} = \frac{7}{8}$

      b. Events A and B are not mutually exclusive because B contains A $\Rightarrow$ $A \cap B = A$, which is not the empty set.

4.9 Because $P(A|B) = \frac{3}{7} \neq \frac{3}{8} = P(A) \Rightarrow$ A and B are not independent.

Because $P(A|C) = 0 \neq \frac{3}{8} = P(A) \Rightarrow$ A and C are not independent.

Because $P(B|C) = 0 \neq \frac{7}{8} = P(B) \Rightarrow$ B and C are not independent.

4.11 $A \cap B = \{6\}; P(A \cap B) = \frac{1}{6} \neq (\frac{1}{6})(\frac{3}{6}) = P(A)P(B)$, thus A and B are not independent.

$A \cap C = \{6\}; P(A \cap C) = \frac{1}{6} \neq (\frac{1}{6})(\frac{4}{6}) = P(A)P(C)$, thus A and C are not independent.

$A \cap D = \{6\}; P(A \cap D) = \frac{1}{6} \neq (\frac{1}{6})(\frac{2}{6}) = P(A)P(D)$, thus A and D are not independent.

$B \cap C = \{4,6\}; P(B \cap C) = \frac{2}{6} = \frac{1}{3} = (\frac{1}{2})(\frac{2}{3}) = P(B)P(C)$, thus B and C are independent.

$B \cap D = \{4,6\}; P(B \cap D) = \frac{2}{6} \neq (\frac{3}{6})(\frac{2}{6}) = P(B)P(D)$, thus B and D are not independent.

$C \cap D = \{4,6\}; P(C \cap D) = \frac{2}{6} \neq (\frac{4}{6})(\frac{2}{6}) = P(C)P(D)$, thus C and D are not independent.

None of the pairs have empty intersections, therefore none of the events are mutually exclusive.

4.13 No, since A and B are not mutually exclusive. Also, $P(A \cap B) = 0.30 \neq 0$

4.15   a. $\overline{A}$: Generator 1 does not work

    b. $B|A$: Generator 2 does not work given that Generator 1 does not work

    c. $A \cup B$: Generator 1 works or Generator 2 works or Both Generators work

4.17   a. $S = \{F_1F_2, F_1F_3, F_1F_4, F_1F_5, F_2F_1, F_2F_3, F_2F_4, F_2F_5, F_3F_1, F_3F_2,$
        $F_3F_4, F_3F_5, F_4F_1, F_4F_2, F_4F_3, F_4F_5, F_5F_1, F_5F_2, F_5F_3, F_5F_4\}$

    b. Let T_1 be the event that the 1st firm chosen is stable and T_2 be the event that the 2nd firm chosen is stable.

$P(T_1) = \frac{3}{5}$ and $P(\overline{T_1}) = \frac{2}{5}$

$P(\text{Both Stable}) = P(T_1 \cap T_2) = P(T_2|T_1)P(T_1) = (\frac{2}{4})(\frac{3}{5}) = \frac{6}{20} = 0.30$

Alternatively, if we designated F_1 and F_2 as the Shakey firms, then we could go to the list of 20 possible outcomes and identify 6 pairs containing just Stable firms: $F_3F_4, F_3F_5, F_4F_3, F_4F_5, F_5F_3, F_5F_4$. Thus, the probability that both firms are Stable is $\frac{6}{20}$

    c. $P(\text{One of two firms is Shakey})$

        $= P(\text{1st chosen is Shakey and 2nd chosen is Stable})$

        $+ P(\text{1st chosen is Stable and 2nd chosen is Shakey})$

        $= P(\overline{T_1} \cap T_2) + P(T_1 \cap \overline{T_2}) = P(T_2|\overline{T_1})P(\overline{T_1}) + P(\overline{T_2}|T_1)P(T_1)$

        $= (\frac{3}{4})(\frac{2}{5}) + (\frac{2}{4})(\frac{3}{5}) = \frac{12}{20} = 0.60$

Alternatively, in the list of 20 outcomes there are 12 pairs which consist of exactly 1 Shakey firm (F_1 or F_2) and exactly 1 Stable firm (F_3 or F_4 or F_5). Thus, the probability of exactly 1 Shakey firm is $\frac{12}{20}$.

    d. $P(\text{Both Shakey}) = P(\overline{T_1} \cap \overline{T_2}) = P(\overline{T_2}|\overline{T_1})P(\overline{T_1}) = (\frac{1}{4})(\frac{2}{5}) = \frac{2}{20} = 0.10$

Alternatively, in the list of 20 outcomes there are 2 pairs in which both firms are Shakey (F_1 or F_2). Thus, the probability that both firms are Shakey is $\frac{2}{20}$.

4.19   a. $P(B|A) = 48/192 = 0.25 \neq 0.291 = P(B)$; thus A and B are dependent.

b. $P(B|A) = 48/192 = 0.25; P(B|A) = 80/248 = 0.323 \Rightarrow \quad P(B|A) \neq P(B|A)$

4.21   a. P(Both customers pay in full) = $(0.70)(0.70) = 0.49$

      b. P(At least one of 2 customers pay in full) = 1 - P(neither customer pays in full) = 1 - $(1-0.70)(1-0.70) = 1 - (0.30)^2 = 0.91$

4.5 Bayes Formula

4.23 Let D be the event loan is defaulted, R_1 applicant is poor risk, R_2 fair risk, and R_3 good risk.

$$P(D) = 0.01, \quad P(R_1|D) = 0.30, \quad P(R_2|D) = 0.40, \quad P(R_3|D) = 0.30$$

$$P(\overline{D}) = 0.01, \quad P(R_1|\overline{D}) = 0.10, \quad P(R_2|\overline{D}) = 0.40, \quad P(R_3|\overline{D}) = 0.50$$

$$P(D|R_1) = \frac{P(R_1|D)P(D)}{P(R_1|D)P(D)+P(R_1|\overline{D})P(\overline{D})} = \frac{(0.30)(0.01)}{(0.30)(0.01)+(0.10)(0.99)} = 0.0294$$

4.25 $P(D_1|A_1) = \frac{P(A_1|D_1)P(D_1)}{P(A_1|D_1)P(D_1)+P(A_1|D_2)P(D_2)+P(A_1|D_3)P(D_3)+P(A_1|D_4)P(D_4)}$

$$= \frac{(0.90)(0.028)}{(0.90)(0.028)+(0.06)(0.012)+(0.02)(0.032)+(0.02)(0.928)} = 0.55851$$

$$P(D_2|A_2) = \frac{(0.80)(0.012)}{(0.05)(0.028)+(0.80)(0.012)+(0.06)(0.032)+(0.01)(0.928)} = 0.43243$$

$$P(D_3|A_3) = \frac{(0.82)(0.032)}{(0.03)(0.028)+(0.05)(0.012)+(0.82)(0.032)+(0.02)(0.928)} = 0.56747$$

4.27 Let F be the event fire occurs and T_i be the event a type i furnace is in the home for $i = 1, 2, 3, 4$, where T_4 represent other types.

$$P(T_1|F) = \frac{P(F|T_1)P(T_1)}{P(F|T_1)P(T_1)+P(F|T_1)P(T_1)+P(F|T_1)P(T_1)+P(F|T_1)P(T_1)}$$

$$= \frac{(0.05)(0.30)}{(0.05)(0.30)+(0.03)(0.25)+(0.02)(0.15)+(0.04)(0.30)} = 0.40$$

4.29 $P(A_2|B_1) = \frac{(0.17)(0.15)}{(0.08)(0.25)+(0.17)(0.15)+(0.10)(0.12)} = 0.4435$

$$P(A_2|B_2) = \frac{(0.12)(0.15)}{(0.18)(0.25)+(0.12)(0.15)+(0.14)(0.12)} = 0.2235$$

$$P(A_2|B_3) = \frac{(0.07)(0.15)}{(0.06)(0.25)+(0.07)(0.15)+(0.08)(0.12)} = 0.2991$$

$$P(A_2|B_4) = \frac{(0.64)(0.15)}{(0.68)(0.25)+(0.64)(0.15)+(0.68)(0.12)} = 0.2762$$

4.8 A Useful Discrete Random Variable: The Binomial

4.31   a. $P(y = 3) = \binom{10}{3}(.2)^3(.8)^7 = 0.201$

         b. $P(y = 2) = \binom{4}{2}(.4)^2(.6)^2 = 0.3456$

         c. $P(y = 12) = \binom{16}{12}(.7)^{12}(.3)^4 = 0.204$

4.33   a. Bar graph of P(y)

         b. $P(y \leq 2) = P(y = 0) + P(y = 1) + P(y = 2) = 0.5$

         c. $P(y \geq 7) = P(y = 7) + P(y = 8) + P(y = 9) + P(y = 10) = 0.13$

         d. $P(1 \leq y \leq 5) = P(y = 1) + P(y = 2) + P(y = 3) + P(y = 4) + P(y = 5) = 0.71$

4.39 Binomial experiment with n=10 and $\pi = 0.60$.

         a. $P(y = 0) = 0.0001$

         b. $P(y = 6) = 0.2508$

         c. $P(y \geq 6) = 1 - P(y < 6) = 1 - (P(0) + P(1) + P(2) + P(3) + P(4) + P(5))$

            $= 1 - (0.3669) = 0.6331$

         d. $P(y = 10) = 0.0060$

4.41 $P(y = 0) = \binom{4}{0}(.8)^0(.2)^4 = 0.0016$

        $P(y = 1) = \binom{4}{1}(.8)^1(.2)^3 = 0.0256$

        $P(y \leq 1) = P(y = 0) + P(y = 1) = 0.0272$

4.45 $\mu = n\pi = (1000)(.5) = 500 \quad \sigma = \sqrt{(1000)(.5)(.5)} = 15.81$

        $\mu \pm 3\sigma = (452.57, 547.43)$

4.47   a. $P(y = 2) = \binom{10}{2}(.1)^2(.9)^8 = 0.1937$

         b. $P(y \geq 2) = 1 - P(y = 0) - P(y = 1) = 1 - \binom{10}{0}(.1)^0(.9)^{10} - \binom{10}{1}(.1)^1(.9)^9 = 0.2639$

         c. $\pi = $ P(either outstanding or good) $= 0.85$; $P(y = 8) = \binom{10}{8}(0.85)^8(.15)^2 = 0.2759$

         d. $\pi = $ P(unsatisfactory) $= 0.05$; $\quad P(y = 0) = \binom{10}{0}(0.05)^0(.95)^{10} = 0.5987$

4.49 Binomial with $n = 15; \pi = .12$

         a. $P(y = 0) = \binom{15}{0}(.12)^0(.88)^{15} = 0.1470$

         b. $P(y \geq 1) = 1 - P(y = 0) = 0.8530$

         c. $P(y \geq 2) = 1 - P(y = 0) - P(y = 1) = 1 - 0.1470 - 0.3006 = 0.5524$

4.51 Binomial with $n = 50; \pi = .17$

         a. $P(y \leq 3) = \sum_{i=0}^{3} \binom{50}{i}(0.07)^i(0.93)^{50-i} = 0.5327$ (using a computer program)

b. The posting of price changes are independent with the same probability 0.07 of being posted incorrectly.

4.9 Probability Distributions for Continuous Random Variables

4.53   a. 0.7580 - 0.5000 = 0.2580

      b. 0.5000 - 0.1151 = 0.3849

4.55   a. 0.9115 - 0.4168 = 0.4947

      b. 0.8849 - 0.6443 = 0.2406

4.57   1-.9599=0.0401

4.59 $z_o = 0$

4.61 $z_o = 2.37$

4.63 $z_o = 1.96$

4.65   a. $P(500 < y < 696) = P(\frac{500-500}{100} < z < \frac{696-500}{100})$
$$= P(0 < z < 1.96) = 0.9750 - 0.5000 = 0.475$$

      b. $P(y > 696) = P(z > \frac{696-500}{100}) = P(z > 1.96) = 1 - 0.9750 = 0.025$

      c. $P(304 < y < 696) = P(\frac{304-500}{100} < z < \frac{696-500}{100})$
$$= P(-1.96 < z < 1.96) = 0.9750 - 0.0250 = 0.95$$

      d. $P(500 - k < y < 500 + k) = P(\frac{500-k-500}{100} < z < \frac{500+k-500}{100})$
$$= P(-.01k < z < .01k) = 0.60.$$
From Table 1 we find,
$$P(-.845 < z < .845) = 0.60. \text{ Thus, } .01k = .845 \Rightarrow k = 84.5$$

4.67   a. 0.9332 - 0.5000 = 0.4332

      b. 0.9641 - 0.5000 = 0.4641

4.71   a. $P(z > 1.96) = 1 - 0.9750 = 0.025$

      b. $P(z > 2.21) = 1 - 0.9864 = 0.0136$

      c. $P(z > 2.86) = 1 - 0.9979 = 0.0021$

      d. $P(z > 0.73) = 1 = 0.7673 = 0.2327$

4.73   a. $\mu = 39; \sigma = 6; P(y > 50) = P(z > \frac{50-39}{6}) = P(z > 1.83) = 1 - 0.9664 = 0.0336$

      b. Since 55 is $\frac{55-39}{6} = 2.67$ std. dev. above $\mu = 39$, thus $P(y > 55) = P(z > 2.67)$
$= 0.0038$. We would then conclude that the voucher has been lost.

4.75 $\mu = 150$; $\sigma = 35$

    a. $P(y > 200) = P(z > \frac{200-150}{35}) = P(z > 1.43) = 0.0764$

    b. $P(y > 220) = P(z > \frac{220-150}{35}) = P(z > 2) = 0.0228$

    c. $P(y < 120) = P(z < \frac{120-150}{35}) = P(z < -0.86) = 0.1949$

    d. $P(100 < y < 200) = P(\frac{100-150}{35} < z < \frac{200-150}{35}) = P(-1.43 < z < 1.43) = 0.8472$

4.77   a. $P(y < 38) = P(z < \frac{38-50}{10}) = P(z < -1.2) = 0.1151$ $\Rightarrow$11.51 percentile.

    b. $P(z \leq k) = 0.67$; $\Rightarrow z = 0.44$

4.11 Random Sampling

4.79 A sample is a random sample if every possible sample of size n from the population has an equal probability of being selected.

4.81 We would number the people in the population from 1 to 1,000 and then go to Table 13. Starting at Line 1, Column 25, we obtain 816, 309, 763, 078, 061, 277, 988, 188, 174, 530, 709, 496, 889, 482, 772. These would be the items in the sample.

4.83 In order to make the sampling random, the network might choose voters based on draws from a random number table, or more simply choose every nth person exiting.

4.85 We would then select the women numbered: 054, 636, 533, 482, 526.

4.12 Sampling Distributions

4.87 The sampling distribution would have a mean of 60 and a standard deviation of $\frac{5}{\sqrt{16}} = 1.25$. If the population distribution is somewhat mound-shaped then the sampling distribution of $\bar{y}$ should be approximately mound-shaped. In this situation, we would expect approximately 95% of the possible values of $\bar{y}$ is lie in $60 \pm (2)(1.25) = (57.5, 62.5)$.

4.89 The following Minitab Commands will provide 500 random samples of 16 observations each from a normal distribution with $\mu = 60, \sigma = 5$:

click on **Calc.**

click on **Random Data**

click on **Normal**

Type **500** in the **Generate rows of data** box

Type **c1-c16** in the **Store in Columns** box

Type **60** in the **Mean** box

Type **5** in the **Standard Deviation** box

click on **Calc.**

click on **Row Statistics**

click on **Mean**

Type **c1-c16** in the **Input** box

Type **c20** in the **Store Results in:** box

click on **Stat.**

click on **Basic Statistics**

click on **Display Descriptive Statistics**

Type **c20** in the **Variables** box

click on **Graph**

click on **Histogram of Data with Normal Curve**

These commands yield a histogram with a normal curve superimposed and the following summary statistcs:

Variable	N	Mean	Median	TrMean	StDev	SE Mean
C20	500	59.982	60.033	59.980	1.215	0.054

Variable	Minimum	Maximum	Q1	Q3
C20	56.340	64.805	59.125	60.818

4.91 a. $P(z < 1.28) = 0.90 \Rightarrow y_{.9} = 930 + (1.28)(130) = 1096.4$

  b. $P(z < -0.6745) = P(z > 0.6745) = 0.25 \Rightarrow$

  $y_{.25} = 930 + (-0.6745)(130) = 842.31; \quad y_{.75} = 930 + (0.6745)(130) = 1017.69 \Rightarrow$
  $IQR = 1017.69 - 842.31 = 175.38$

4.93 $P(z > 1.645) = 0.05 \Rightarrow k = 125 + (1.645)(32) = 177.64 \Rightarrow$

  Facility size should be at least 178.

  $P(z > 2.326) = 0.01 \Rightarrow k = 125 + (2.326)(32) = 199.4 \Rightarrow$

  Facility size should be at least 200.

4.95 $\mu = 2.1; \quad \sigma = 0.3$

  a. $P(y > 2.7) = P(z > \frac{2.7-2.1}{0.3}) = P(z > 2) = 0.0228$

  b. $P(z > 0.6745) = 0.25 \Rightarrow y_{.75} = 2.1 + (0.6745)(0.3) = 2.30$

  c. Let μ_N be the new value of the mean. We need $P(y > 2.7) \le 0.05$.
  From Table 1, $0.05 = P(z > 1.645)$ and $0.05 = P(y > 2.7) = P(\frac{y - \mu_N}{.3} > \frac{2.7 - \mu_N}{0.3}) \Rightarrow$
  $\frac{2.7 - \mu_N}{0.3} = 1.645 \Rightarrow \mu_N = 2.7 - (0.3)(1.645) = 2.2065$

4.97 Individual baggage weight has $\mu = 95; \sigma = 35;$ Total weight has mean $n\mu = (200)(95) = 19,000;$ and standard deviation $\sqrt{n}\sigma = \sqrt{200}(35) = 494.97.$ Therefore, $P(y > 20,000) = P(z > \frac{20,000 - 19,000}{494.97}) = P(z > 2.02) = 0.0217$

Supplementary Exercises

4.99 No. The last date may not be representative of all days in the month.

4.101 a. $P(y < 5) =$
  $= \binom{20}{0}(.5)^0(.5)^{20} + \binom{20}{1}(.5)^1(.5)^{19} + \binom{20}{2}(.5)^2(.5)^{18} + \binom{20}{3}(.5)^3(.5)^{17} + \binom{20}{4}(.5)^4(.5)^{16}$
  $= 0.0059$
  $\mu = n\pi = (20)(0.5) = 10; \sigma = \sqrt{20)(0.5)(0.5)} = 2.236;$
  $P(y < 5) \approx P(z \le \frac{5-10}{2.236}) = P(z < -2.24) = 0.0125$

b. For the binomial distribution, $P(y < 5) = P(y \leq 4)$.

Thus, we have $P(y < 5) = P(y \leq 4) \approx P(z \leq \frac{4.5-10}{2.236}) = P(z < -2.46) = 0.0069$

c. $P(8 < y < 14) = P(9 \leq y \leq 13)$

$= \binom{20}{9}(.5)^9(.5)^{11} + \binom{20}{10}(.5)^{10}(.5)^{10} + \binom{20}{11}(.5)^{11}(.5)^9 + \binom{20}{12}(.5)^{12}(.5)^8 + \binom{20}{13}(.5)^{13}(.5)^7$

$= 0.6906$

Using normal approximation with correction: $P(8 < y < 14) = P(9 \leq y \leq 13)$

$= P(y \leq 13) - P(y \leq 8) \approx P(z \leq \frac{13.5-10}{2.236}) - P(z \leq \frac{8.5-10}{2.236})$

$= P(z < 1.57) - P(z < -0.67) = 0.6904$

4.103 $P(4 \leq y \leq 6) = P(z < \frac{6.5-5}{1.58}) - P(z < \frac{3.5-5}{1.58}) = P(z < .95) - P(z < -.95) = 0.6579$.

With the correction the approximation is very accurate.

4.105   a. $P(y > 2265) = P(z > \frac{2265-2250}{10.2}) = P(Z > 1.47) = 0.0707$

b. Approximately normal with mean = 2250 and standard deviation = $\frac{10.2}{\sqrt{15}} = 2.63$

4.107 $P(\bar{y} \geq 2268) = P(z \geq \frac{2268-2250}{10.2/\sqrt{15}}) = P(z \geq 6.84) \approx 0$.

4.109   a. The normal probability plot and box plots are given here.

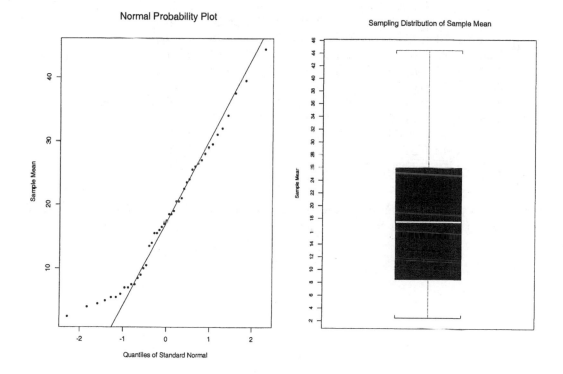

Note that the plotted points deviate from the straight-line.

b. Since the population distribution is skewed to the right for this sample and with the sample size only 2, the sampling distribution for $\bar{y}$ will be skewed to the right.

4.113 a. The random variable y is the number of people (out of 1,000) who responded that they were planning to buy the reformulated drink.

b. Y is a binomial variable and so has mean $n\pi$ and variance $n\pi(1-\pi)$. We do not know the value of the parameter π; so we cannot compute the true mean and variance. We can only estimate them by using the sample proportion as an estimate.

c. Using the sample estimate of π, we can compute $P(y \leq 250)$ by the normal approximation if $n\pi$ and $n(1-\pi)$ are both greater than 5.

4.115 $n = 400 \qquad \pi = 0.2$

a. $\mu = n\pi = 400(.20) = 80; \sigma = \sqrt{400(.2)(.8)} = 8$
$P(y \leq 25) \approx P(z \leq \frac{25-80}{8}) = P(z \leq -6.875) \approx 0.$

b. The ad is not successful. With $\pi = .20$, we expect 80 positive responses out of 400 but we observed only 25. The probability of getting so few positive responses is virtually 0 if $\pi = .20$. We therefore conclude that π is much less than 0.20.

4.117 The sampling distribution of the sample mean consists of the following values for $\bar{y}$ and their frequency of occurrence.

$\bar{y}$	$P(\bar{y})$	$\bar{y}$	$P(\bar{y})$	$\bar{y}$	$P(\bar{y})$	$\bar{y}$	$P(\bar{y})$	$\bar{y}$	$P(\bar{y})$
7.25	1/70	17.00	1/70	21.50	2/70	26.00	1/70	35.75	1/70
10.50	1/70	17.25	1/70	21.75	1/70	26.25	1/70		
11.25	1/70	17.50	1/70	22.25	2/70	26.50	1/70		
11.50	1/70	18.00	2/70	22.50	2/70	26.75	1/70		
12.00	1/70	18.25	1/70	23.25	2/70	27.50	2/70		
12.25	1/70	18.75	2/70	23.50	1/70	28.25	1/70		
13.00	2/70	19.00	1/70	23.75	1/70	29.00	2/70		
14.00	2/70	19.25	1/70	24.00	1/70	30.00	2/70		
14.75	1/70	19.50	1/70	24.25	2/70	30.75	1/70		
15.50	2/70	19.75	2/70	24.75	1/70	31.00	1/70		
16.25	1/70	20.50	2/70	25.00	2/70	31.50	1/70		
16.50	1/70	20.75	2/70	25.50	1/70	31.75	1/70		
16.75	1/70	21.25	1/70	25.75	1/70	32.50	1/70		

4.119 The sampling distribution of the sample median consists of the following values for the median (M) and their frequency of occurrence.

M	P(M)
7.5	5/70
9.0	4/70
10.5	8/70
15.5	3/70
17.0	6/70
17.5	2/70
18.5	9/70
19.0	4/70
20.5	6/70
22.5	1/70
24.0	2/70
25.5	3/70
27.0	8/70
32.0	4/70
34.0	5/70

4.120   a. The normal probability plot is given here.

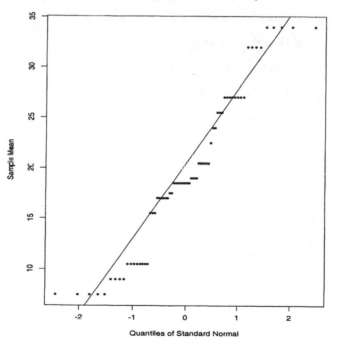

Note that the plotted points deviate considerably from the straight-line. Thus, the sampling distribution is not approximated very well by a normal distribution. If the sample size was much larger than 4 the approximation would be greatly improved.

b. The population median equals $\frac{12+25}{2} = 18.5$ whereas the mean of the 70 values of the sample median is 19.536. The values differ by a significant amount due to the fact that the sample size was only 4.

4.121 a.,b. The mean and standard deviation of the sampling distribution of $\bar{y}$ are given when the population distribution has values $\mu = 100, \sigma = 15$:

Sample Size	Mean	Standard Deviation
5	100	6.708
20	100	3.354
80	100	1.677

c. As the sample size increases, the sampling distribution of $\bar{y}$ concentrates about the true value of μ. For $n = 5$ and 20, the values of $\bar{y}$ could be a considerable distance from 100.

4.123 $n = 36, \mu = 40, \sigma = 12$

a. The sampling distribution of $\bar{y}$ is approximately normal with a mean of 40 and a standard deviation of $\frac{12}{\sqrt{36}} = 2$

b. $P(\bar{y} > 36) = P(z > \frac{36-40}{2}) = P(z > -2) = 0.9772$

c. $P(\bar{y} < 30) = P(z < \frac{30-40}{2}) = P(z < -5) = 2.87 x 10^{-7}$

d. $P(z > 1.645) = 0.05 \Rightarrow k = 40 + (1.645)(2) = 43.29$

4.125 a. Mean = 10, Standard Deviation = $\frac{10}{\sqrt{25}} = 2$

b. Mean = 10, Standard Deviation = $\frac{10}{\sqrt{100}} = 1$

c. Mean = 10, Standard Deviation = $\frac{20}{\sqrt{25}} = 4$

d. Mean = 10, Standard Deviation = $\frac{20}{\sqrt{100}} = 2$

Chapter 5: Inferences about Population Central Values

5.1 Introduction and Case Study

5.1   a. All registered voters in the state.

       b. Simple random sample from a list of registered voters.

5.3   We might think that the actual average lifetime is less than the proposed 1500 hours.

       a. The population is the lifetime of all fuses produced by the manufacturer during a selected period of time.

       b. Testing a hypothesis.

5.2 Estimation of μ

5.5   a. $12.3 \pm (1.96)\frac{0.2}{\sqrt{25}} = (12.22, 12.38)$

       b. We are 95% confident that the average weight of a box of corn flakes is between 12.22 and 12.38 oz.

5.7   a. 36

       b. 7

       c. 7

5.9   a. $5.2 \pm (2.58)(\frac{7.5}{\sqrt{10}}) = 5.2 \pm 6.12 = (-.92, 11.32)$

       b. Since the sample size is small, the condition that the distribution of profit margins needs to be normal is crucial. Similarly, with n=10, replacing σ with s is questionable. Section 5.7 will provide more details about this type of situation.

5.11 $3.2 \pm (1.96)(\frac{1.1}{\sqrt{150}}) = 3.2 \pm 0.18 = (3.02, 3.38)$

5.13 $850 \pm (1.96)(\frac{100}{\sqrt{60}}) = 850 \pm 25.3 = (824.7, 875.3)$

5.15 $430 \pm (1.96)(\frac{262}{\sqrt{900}}) = 430 \pm 17.1 = (412.9, 447.1)$

5.17 $0.18 \pm (1.96)(\frac{0.08}{\sqrt{100}}) = 0.18 \pm 0.016 = (0.164, 0.196)$

We are 95% confident that the mean protein content for the population is between 0.164 and 0.196.

5.3 Choosing the Sample Size for Estimating μ

5.19 a. n decreases

    b. n increases

    c. n increases

5.21 $\hat{\sigma} = 13, E = 3, \alpha = .01 \Rightarrow n = \frac{(2.58)^2 (13)^2}{(3)^2} = 125$

5.23 a. $\hat{\sigma} = \frac{1500-400}{4} = 325 \Rightarrow n = \frac{(2.58)^2 (325)^2}{(25)^2} = 1125$

    b. The 95% level of confidence implies that there will be a 1 in 20 chance, over a large number of samples, that the confidence interval will not contain the population average rent. The 99% level of confidence implies there is only a 1 in 100 chance of not containing the average. Thus, we would increase the odds of not containing the true average five-fold.

5.5 Choosing the Sample Size for μ

5.25 a. $n = \frac{(.225)^2 (1.96+2.33)^2}{(.1)^2} = 94$

    b. The sample size is somewhat larger.

5.27 a.-c. The power values are given here:

n	α	μ_a	39	40	41	42	43	44
50	0.05	$PWR(\mu_a)$	0.3511	0.8107	0.9840	0.9997	1.0000	1.0000
50	0.025	$PWR(\mu_a)$	0.2428	0.7141	0.9662	0.9990	0.9999	1.0000
20	0.05	$PWR(\mu_a)$	0.1987	0.4809	0.7736	0.9394	0.9906	0.9992

The power curves are plotted here:

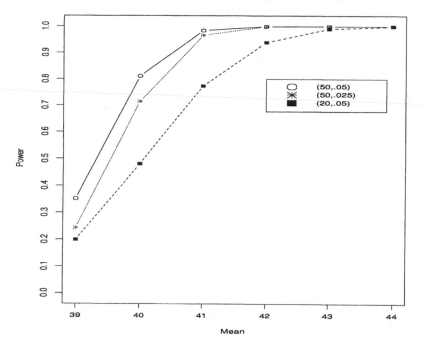

Power Curves for Three Situations

5.29 a. Simulate data using computer program

b. $PWR(41.5) = P(z > 1.645 - \frac{|40-41.5|}{8/\sqrt{16}}) = P(z > .895) = 0.1854$

c. Expected Number is $(100)(.1854) = 18.54 \approx 19$

d. For $\mu = 38, PWR(38) = P(z > 1.645 - \frac{|40-38|}{8/\sqrt{16}}) = P(z > .645) = 0.2595$

Expected Number is $(100)(.2595) = 25.95 \approx 26$

For $\mu = 43, PWR(43) = P(z > 1.645 - \frac{|40-43|}{8/\sqrt{16}}) = P(z > .145) = 0.4424$

Expected Number is $(100)(.4424) = 44.24 \approx 44$

5.31 $H_o : \mu \geq 16$ vs $H_a : \mu < 16$

$\alpha = 0.05, \beta = 0.10$, whenever $\mu \leq 12, \sigma = 7.64$

$z_{0.05} = 1.645, z_{0.10} = 1.28, n = \frac{(7.64)^2(1.645+1.28)^2}{(12-16)^2} = 31.2 \Rightarrow n = 32$

5.33 The power values are computed using the following formulas:

$n = 90, \alpha = 0.05, PWR(\mu_a) = P(z > 1.645 - \frac{|2-\mu_a|}{1.05/\sqrt{90}})$

$n = 90, \alpha = 0.01, PWR(\mu_a) = P(z > 2.33 - \frac{|2-\mu_a|}{1.05/\sqrt{90}})$

$n = 50, \alpha = 0.05, PWR(\mu_a) = P(z > 1.645 - \frac{|2-\mu_a|}{1.05/\sqrt{50}})$

The power values are given here:

n	α	μ_a	2.1	2.2	2.3	2.4	2.5
90	0.05	$PWR(\mu_a)$	0.2292	0.5644	0.8567	0.9755	0.9980
90	0.01	$PWR(\mu_a)$	0.0769	0.3005	0.6482	0.9004	0.9857
50	0.05	$PWR(\mu_a)$	0.1656	0.3828	0.6463	0.8529	0.9575

The power curves are plotted here:

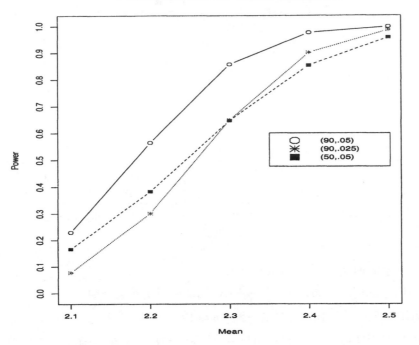

a. Reducing α from 0.05 to 0.01 reduces the power of the test.

b. Reducing the sample size from 90 to 50 reduces the power of the test.

5.35 $z = \frac{542-525}{76/\sqrt{100}} = 2.24 > 1.645 = z_{0.05} \Rightarrow$ Reject H_o.

There is sufficient evidence to conclude that the mean has been increased above 525.

5.37 a. $z = \frac{7.3-5}{4.6/\sqrt{35}} = 2.96 > 1.645 = z_{0.05} \Rightarrow$ Reject H_o.

There is sufficient evidence to conclude that the mean weight reduction is greater than 5 pounds.

b. Probability of Type I error is 0.05.

Probability of Type II error is 0 since H_o was rejected.

5.39 $H_o : \mu \le 0.3$ vs $H_a : \mu > 0.3$

$\alpha = 0.05, n = 60, \bar{y} = 0.7, s = 0.4$

a. $z = \frac{0.7-0.3}{0.4/\sqrt{60}} = 7.75 > 1.645 = z_{0.05} \Rightarrow$ Reject H_o.

There is sufficient evidence to conclude that the mean number of newly decayed teeth exceeds 0.3.

b. Since we are interested in testing $\mu > 0.3$, a two-tailed test would not be appropriate.

5.6 The Level of Significance of a Statistical Test

5.41 p-value $= 0.0359 > 0.025 = \alpha \Rightarrow$

No, there is not significant evidence that the mean is greater than 45. With $\alpha = 0.025$, the researcher is demanding greater evidence in the data to support the research hypothesis.

5.43 Yes, since p-value would now equal $0.0164/2 = 0.0082$, which is less than 0.01, the value of α.

5.45 p-value $= P(z \geq \frac{13.5-14}{3.62/\sqrt{40}}) = P(z \geq -0.87) = 0.8078 > 0.05 = \alpha \Rightarrow$

No, there is not significant evidence that the mean is greater than 14. This is the opposite of the conclusion reached in 5.44. This emphasizes the importance of selecting the research hypothesis, H_a.

5.47 $H_o : \mu = 1.6$ versus $H_a : \mu \neq 1.6$,

$n = 36, \bar{y} = 2.2, s = 0.57, \alpha = 0.05$.

p-value $= 2P(z \geq \frac{|2.2-1.6|}{.57/\sqrt{36}}) = 2P(z \geq 6.32) < 0.0001 < 0.05 = \alpha \Rightarrow$

Yes, there is significant evidence that the mean time delay differs from 1.6 seconds.

5.7 Inferences about μ for a Normal Population, σ Unknown

5.49   a. Reject H_o if $t \leq -1.761$

b. Reject H_o if $|t| \geq 2.074$

c. Reject H_o if $t \geq 2.015$

5.51 Reject H_o if $t \geq t_{.05,17} = 1.740$

$t = \frac{16.2-15}{3.1/\sqrt{18}} = 1.64 \Rightarrow$ Fail to reject H_o and conclude data does not support the hypothesis that the mean is greater than 15.

5.53 $H_o : \mu \le 80$ versus $H_a : \mu > 80, n = 20, \bar{y} = 82.05, s = 10.88$ $\qquad$ Reject H_o if $t \ge 1.729$.

$t = \frac{82.05 - 80}{10.88/\sqrt{20}} = 0.84 \Rightarrow$

Fail to reject H_o and conclude data does not support the hypothesis that the mean reading comprehension is greater than 80.

The level of significance is given by p-value $= P(t \ge 0.84) \approx 0.20 > 0.05 = \alpha$.

5.55 $n = 15, \bar{y} = 31.47, s = 5.04$

    a. $31.47 \pm (2.977)(5.04)/\sqrt{15} \Rightarrow 31.47 \pm 3.87 \Rightarrow (27600, 35340)$ is a 99% C.I. on the mean miles driven.

    b. $H_o : \mu \ge 35$ versus $H_a : \mu < 35$ $\qquad$ Reject H_o if $t \le -2.624$.

    $t = \frac{31.47 - 35}{5.04/\sqrt{15}} = -2.71 \Rightarrow$

    Reject H_o and conclude data supports the hypothesis that the mean miles driven is less than 35,000 miles.

    Level of significance is given by p-value $= P(t \le -2.71) \Rightarrow 0.005 < p - value < 0.01$.

5.57   a. $4.95 \pm (2.365)(0.45)/\sqrt{8} \Rightarrow 4.95 \pm 0.38 \Rightarrow (4.57, 5.33)$ is a 95% C.I. on the mean dissolved oxygen level.

    b. There is inconclusive evidence that the mean is less than 5 since the C.I. contains values both less and greater than 5.

    c. $H_o : \mu \ge 5$ versus $H_a : \mu < 5$, $t = \frac{4.95 - 5}{.45/\sqrt{8}} = -0.31 \Rightarrow$ p-value $= P(t \le -0.31) = P(t \ge 0.31) \Rightarrow 0.25 < p - value < 0.40$ (Using a computer program p-value $= 0.3828$). Fail to reject H_o and conclude the data does not support that the mean is less than 5.

5.59   a. Untreated: $43.6 \pm (1.833)(5.7)/\sqrt{10} \Rightarrow (40.3, 46.9)$

    Treated: $36.1 \pm (1.833)(4.9)/\sqrt{10} \Rightarrow (33.3, 38.9)$

    We are 90% confident that the average height of untreated shrups is between 40.3 cm and 46.9 cm. We are 90% confident that the average height of treated shrups is between 33.3 cm and 38.9 cm.

    b. The two intervals do not overlap. This would indicate that the average heights of the treated and untreated shrups are significantly different.

5.60 Before: $23.22 \pm (1.833)(4.25)/\sqrt{10} \Rightarrow (20.76, 25.68)$

After: $25.33 \pm (1.833)(4.25)/\sqrt{10} \Rightarrow (22.87, 27.79)$

Since the two intervals overlap, there is not strong evidence of an increase in the average mpg after installing the device.

5.61   a. $H_o : \mu_{After} - \mu_{Before} \le 0$ versus $H_a : \mu_{After} - \mu_{Before} > 0$ which implies

    $H_o : \mu_{After} \le \mu_{Before}$ versus $H_a : \mu_{After} > \mu_{Before}$

    $t = \frac{2.11 - 0}{7.54/\sqrt{10}} = 0.88 \Rightarrow$ p-value $= P(t \ge 0.88) = 0.2009 > 0.05$

    There is not evidence that the mean change in mpg is greater than 0, i.e., the mean mpg does not appear to be increased after installing the device.

b. $2.11 \pm (1.833)(7.54)/\sqrt{10} \Rightarrow (-2.26, 6.48) \Rightarrow$

Using the decision rule: Reject $H_o : \mu \leq \mu_o$ in favor of $H_a : \mu > \mu_o$ if μ_o is less than the lower limit of the C.I., we have that since 0 is greater than the lower limit of the C.I., -2.26, we fail to reject H_o and conclude that there is not significant evidence that the difference in the average mpg, $\mu_{After} - \mu_{Before}$ is greater than 0.

5.62   a. Let $\mu_C = \mu_{After} - \mu_{Before}$. The probabilities of Type II error are computed using Table 3 on page 1094 with $d = \frac{|\mu_a - 0|}{7.54}$ and are given here:

μ_C	1.0	2.0	3.0	4.0	5.0	6.0	7.0	8.0	9.0
d	0.13	.27	.40	0.53	0.66	.80	.93	1.06	1.19
$\beta(\mu_a)$	0.89	0.81	0.68	0.54	0.39	0.25	0.14	0.07	0.03

The probabilities of Type II error are large for values of μ_C which are of practical importance.

b. Since the probabilities of Type II errors are large, the sample size should be increased. The models, age, condition of the cars used in the study should be considered. The type of driving conditions and experience of drivers are also important factors to be considered in order for the results to be generalizable to a broad population of potential users of the device.

5.8 Inferences about the Median

5.63   a. $L_{.05} = 5, U_{.05} = 16$

      b. $L_{.05} = 6, U_{.05} = 15$

      The intervals are somewhat narrower

5.65 Reject H_o if $B \geq 20$

5.67 Reject H_o if $B \leq 20$ or $B \geq 33$

5.69   a. The normal probability plot is given here:

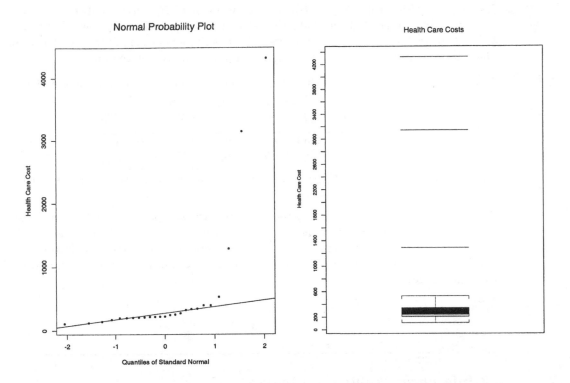

Normal Probability Plot Health Care Costs

The data set does not appear to be a sample from a normal distribution, since a large proportion of the values are outliers as depicted in the box plot and several points are a considerable distance from the line in the normal probability plot. The data appears to be from an extremely right skewed distribution.

b. Because of the skewness, the median would be a better choice than the mean.

c. (207, 345) We are 95% confident that the median amount spent on healthcare by the population of hourly workers is between \$207 and \$345 per year.

d. Reject $H_o : M \le 400$ if $B \ge 25 - 7 = 18$.

We obtain B = 4; Since $4 < 18$, do not reject $H_o : M \le 400$. The data fails to demonstrate that the median amount spend on health care is greater than \$400.

5.72 a. Fund A: 95% C.I. on the Mean: $13.65 \pm (2.262)(15.87)/\sqrt{10} \Rightarrow (2.30, 25.00)$
     Median = 20, 95% C.I. on the Median: $(y_{(1)}, y_{(10)}) \Rightarrow (-8.9, 32.9)$

   Fund B: 95% C.I. on the Mean: $16.56 \pm (2.262)(16.23)/\sqrt{10} \Rightarrow (4.95, 28.17)$
     Median = 16.6, 95% C.I. on the Median: $(y_{(1)}, y_{(10)}) \Rightarrow (-8.4, 41.8)$

   b. The normal probability and box plots are given here:

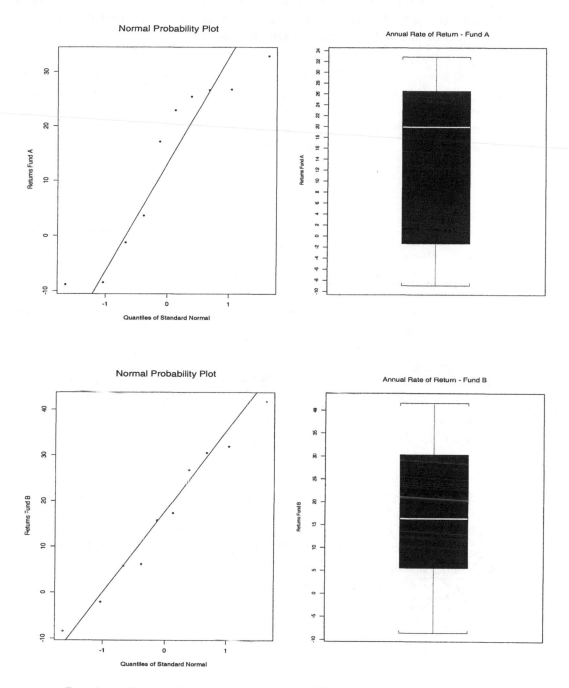

Based on the boxplots and normal probability plots, the median is the most appropriate measure for Fund A and the mean is most appropriate for Fund B.

5.73   a. Using $\alpha = 0.05$; Reject $H_o : M \le 10$ in favor of $H_o : M > 10$ if $B \ge 9$

Fund A: $B = 6 \Rightarrow$ Fail to reject H_o, there is not sufficient evidence to conclude that the median annual rate of returen is greater than 10%.

Fund B: $B = 6 \Rightarrow$ Fail to reject H_o, there is not sufficient evidence to conclude that the median annual rate of returen is greater than 10%.

b. Using $\alpha = 0.05$; Reject H_o if p-value ≤ 0.05.

Fund A: $t = \frac{13.65-10}{15.87/\sqrt{10}} = 0.73 \Rightarrow$ p-value $= P(t \geq 0.73) = 0.24 \Rightarrow$ Fail to reject H_o, there is not sufficient evidence to conclude that the mean annual rate of returen is greater than 10%.

Fund B: $t = \frac{16.56-10}{16.23/\sqrt{10}} = 1.28 \Rightarrow$ p-value $= P(t \geq 1.28) = 0.12 \Rightarrow$ Fail to reject H_o, there is not sufficient evidence to conclude that the mean annual rate of returen is greater than 10%.

Supplementary Exercises

5.77   a. $H_a : \mu \neq 220$

b. Reject H_o if $|z| \geq 1.96$. Since $z = \frac{208-220}{55/\sqrt{100}} = -2.18 \Rightarrow$ Reject H_o and conclude there is significant evidence in the data that the mean has changed from 220.

c. p-value $= 2P(z \geq |-2.18|) = 0.0292$

d. 95% C.I.: $208 \pm (1.96)(55)/\sqrt{100} \Rightarrow (197.22, 218.78)$

5.79   a. $\bar{y} = 1.466$

b. 95% C.I.: $1.466 \pm (2.145)(.3765)/\sqrt{15} \Rightarrow (1.26, 1.67)$

We are 95% confident that the average mercury content after the accident is between 1.28 and $41.66 mg/m^3$

c. $H_a : \mu > 1.20$ Reject H_o if $t \geq 1.761$ $t = \frac{1.466-1.2}{.3765/\sqrt{15}} = 2.74 \Rightarrow$

There is sufficient evidence that the mean mercury concentration has increased.

d. Using Table 3, we obtain the following with $d = \frac{|\mu_a - 1.2|}{.32}$

μ_a	d	$PWR(\mu_a)$
1.28	0.250	0.235
1.32	0.375	0.396
1.36	0.500	0.578
1.40	0.625	0.744

5.81   a. F Type II
       b. T
       c. F α
       d. F decreases

5.83   99% C.I.: $3250 \pm (2.576)(420)/\sqrt{70} \Rightarrow (3120.7, 3379.3)$

5.85 $H_0 : \mu = 80\%$ versus $H_a : \mu \neq 80\%$,
       $n = 30, \bar{y} = 78.3\%, s = 2.9\%$
       p-value $= 2P(t \geq \frac{|78.3-80|}{2.9/\sqrt{30}}) = 2P(t \geq 3.21) = 0.0032 < 0.05 = \alpha.$
       Yes, the data contradicts the manufacturer's claim.

5.87 $H_0 : \mu \geq 2.3$ versus $H_a : \mu < 2.3$,
       $n = 200, \bar{y} = 2.24, s = 0.31, \alpha = 0.05$
       a. p-value $= P(z \leq \frac{2.24-2.3}{0.31/\sqrt{200}}) = P(z \leq -2.74) = 0.0031 < 0.05 = \alpha.$
          Yes, there is sufficient evidence to conclude that the average GPA is less than 2.3.
       b. The observed difference is $2.24 - 2.3 = -0.06$. This difference is so small as to
          have no practical consequence in terms of the difference in student performance
          in the classroom.

5.88   a. Normal probability and box plots are given here:

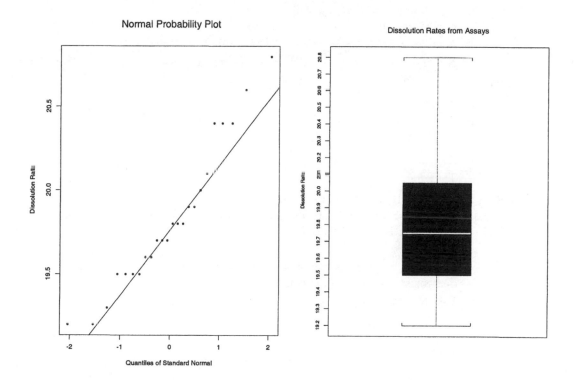

       The plots indicate that the data was selected from a population having a distri-
       bution which is somewhat skewed to the right but only slightly, since there are no
       outliers indicated on the box plot.

41

b. $\bar{y} = 19.83$;   99% C.I.: $19.83 \pm (2.807)(0.43)/\sqrt{24} \Rightarrow (19.58, 20.08)$

c. $H_0 : \mu \geq 20$ versus $H_a : \mu < 20$,
$n = 24, \bar{y} = 19.83, s = 0.43, \alpha = 0.01$
p-value $= P(t \leq \frac{19.83-20}{0.43/\sqrt{24}}) = P(t \leq -1.94) = 0.0324 > 0.01 = \alpha$.
No, there is not sufficient evidence to conclude that the average dissolution rate is less than 20.

d. From Table 3 in Appendix with $d = \frac{|19.6-20|}{.43} = .93, df = 23, \alpha = 0.01$, we obtain $\beta(19.6) = .025$ (actual this value was obtained from a computer program since this degree of accuracy could not be obtained from Table 3).

5.91 a. $\bar{y} = 74.2$;   95% C.I.: $74.2 \pm (2.145)(44.2)/\sqrt{15} \Rightarrow (49.72, 98.68)$

b. $H_0 : \mu \leq 50$ versus $H_a : \mu > 50$,
$n = 15, \alpha = 0.05$
p-value $= P(t \geq \frac{74.2-50}{44.2/\sqrt{15}}) = P(t \geq 2.12) = 0.0262 < 0.05 = \alpha$.
Yes, there is sufficient evidence to conclude that the average daily output is greater than 50 tons of ore.

5.93 a. 95% C.I.: $8.93 \pm (2.201)(0.314)/\sqrt{12} \Rightarrow (8.73, 9.13)$
We are 95% confident that the average potency for the batch is between 8.73 and 9.13.

b. It would depend on the physical characteristics of the vat.

c. $H_0 : \mu = 9$ versus $H_a : \mu \neq 9$,
$n = 12, \alpha = 0.05$
p-value $= 2P(t \geq \frac{|8.93-9|}{0.314/\sqrt{12}}) = 2P(t \geq 0.77) = 0.4562 > 0.05 = \alpha$.
No, there is not sufficient evidence that the mean potency differs from 9.0.

5.95 a. T
b. F
c. T
d. T
e. F
f. T

5.97 a. $H_o : \mu \geq 10$ versus $H_a : \mu < 10$
Data is a SRS from a normally distributed population with $\sigma = 2.4$
Reject H_o is $z \leq -1.645$
$z = \frac{8.8-10}{2.4/\sqrt{16}} = -2$
Reject H_o. There is significant evidence in the data that the new process has decreased the average tar content.

b. Yes; No; Yes; No

5.99 $H_o : \mu \geq 8.2; H_a : \mu < 8.2$
$z = \frac{7.6-8.2}{1.8/\sqrt{50}} = -2.36$
The level of significance is given by p-value $= P(z < -2.36) = 0.0091 < 0.05 = \alpha$.
Thus, reject Ho and conclude that the average stress level during peak hours for drivers under the new system is less than 8.2.

5.101 a. $\bar{y} = 30.514, s = 12.358, n = 35$

95% C.I.: $30.514 \pm (2.032)(12.358)/\sqrt{35} \Rightarrow (26.27, 34.76)$

We are 95% confident that this interval captures the population mean exercise capacity.

b. 99% C.I.: $30.514 \pm (2.728)(12.358)/\sqrt{35} \Rightarrow (24.81, 36.21)$

The 99% C.I. is somewhat wider that the 95% C.I.

5.102 $n = \frac{(12.36)^2(1.96)^2}{(1)^2} = 586.9 \Rightarrow n = 587$

5.105 $n = 30, \bar{y} = 98.4, s = 0.15$

90% C.I. on μ : $98.4 \pm (1.699)(.15)/\sqrt{30} \Rightarrow (98.35, 98.45)$

5.109 $n = 50, \bar{y} = 75, s = 15$

95% C.I. on μ : $75 \pm (2.010)(15)/\sqrt{50} \Rightarrow (70.7, 79.3)$

5.111 $n = 40, \bar{y} = 35, s = 6.3$

95% C.I. on μ : $35 \pm (2.023)(6.3)/\sqrt{40} \Rightarrow (33.0, 37.0)$

5.113 $\hat{\sigma} = \frac{400-40}{4} = 90 \Rightarrow n = \frac{(1.96)^2(90)^2}{(10)^2} = 311.2 \Rightarrow n = 312$

5.115 $\hat{\sigma} = \frac{20000}{4} = 5000 \Rightarrow n = \frac{(1.96)^2(5000)^2}{(750)^2} = 170.7 \Rightarrow n = 171$

5.117 The data are skewed to the right because some patients survive much longer than the majority of the patients. The mean would be a misleading indicator of survivability because a few extreme values in a data set has a strong affect on the mean. In this case, the center as measured by the mean, would be much larger than the vast majority of the survival times.

Chapter 6: Inferences Comparing Two Population Central Values

---

6.2 Inferences about $\mu_1 - \mu_2$ Independent Samples

6.1   a. Reject H_o if $|t| \geq 2.064$

      b. Reject H_o if $t \geq 2.624$

      c. Reject H_o if $t \leq -1.860$

6.5   a. $H_o : \mu_A = \mu_B$ versus $H_a : \mu_A \neq \mu_B$; $p - value = 0.065 \Rightarrow$ The data do not provide sufficient evidence to conclude there is a difference in mean oxygen content.

      b. The separate variance t'-test was used since it has df given by
$$c = \frac{(.157)^2/15}{(.157)^2/15 + (.320)^2/15} = 0.194; \Rightarrow df = \frac{(15-1)(15-1)}{(1-.194)^2(15-1)+(.194)^2(15-1)} \approx 20.$$
If the pooled t-test was used the df=15+15-2=28 but the separate variance test has $df \approx 20$, as is shown on printout.

      c. The Above-town and Below-town data sets appear to be normally distributed, with the exception that the box-plot does display an outlier for the Below-town data. The Below-town data appears to be more variable than the Above-town data. The two data sets consist of independent random samples.

      d. The 95% C.I. estimate for the difference in means is (-0.013, 0.378) ppm. The observed difference (0.183 ppm) is not significant.

6.7 We want to test $H_o : \mu_{No} = \mu_{Sub}$ versus $H_a : \mu_{No} \neq \mu_{Sub}$; $p - value = 0.0049 \Rightarrow$ The data supports the contention that people that receive a daily newspaper have a greater knowledge of current events. The stem-and-leaf plots indicate that both data sets are from normally distributed populations but with different variances. The problem description indicates that the two random samples were independently selected. A 95% C.I. on the mean difference is (-15.5, -2.2) which reflects the lower values from the people who did not read a daily newspaper.

6.9   a. The box plots are given here:

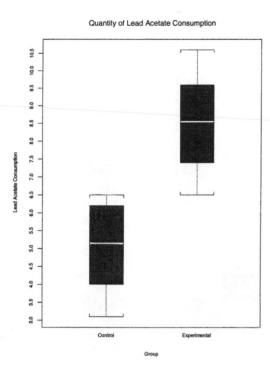

Quantity of Lead Acetate Consumption

Box plots of two data sets indicates that the two data sets are random samples from normally distributed populations with equal variances.

b. $H_o : \mu_C \geq \mu_E$ versus $H_a : \mu_C < \mu_E$;
$t = \frac{5.06 - 8.56}{1.3377\sqrt{\frac{1}{10} + \frac{1}{10}}} = -5.86 \Rightarrow$ with $df = 18, \Rightarrow p - value < 0.0005 \Rightarrow$
Reject H_o and conclude the data provide sufficient evidence that the mean quantity of lead acetate consumed is greater for the experimental group.

A 95% C.I. on the difference in the mean quantity of lead acetate consumed is (-4.76, -2.24), which would indicate that the control group consumes on the average between 4.7 and 2.2 units less than the experimental group.

6.14　a. $H_o : \mu_{96} \geq \mu_{82}$ versus $H_a : \mu_{96} < \mu_{82}$;
$t = \frac{7.00 - 54.30}{\sqrt{\frac{(3.89)^2}{13} + \frac{(15.7)^2}{13}}} = -10.58 \Rightarrow$ with $df = 13, \quad p - value < 0.0005 \Rightarrow$
reject H_o and conclude the data provide sufficient evidence that there has been a significant decrease in mean PCB content.

b. A 95% C.I. on the difference in the mean PCB content of herring gull eggs is (-4.76, -2.24), which would indicate that the decrease in mean PCB content from 1982 to 1996 is between 2.24 and 4.76.

c. The box plots are given here:

45

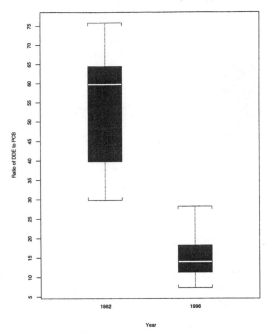

Ratio of DDE to PCB in Herring Gull Eggs

The box plots of the PCB data from the two years both appear to support random samples from normal distributions, although the 1982 data is somewhat skewed to the left. The variances for the two years are substantially different hence the separate variance t-test was applied in a.

    d. Since the data for 1982 and 1996 were collected at the same sites, there may be correlation between the two years. There may be also be spatial correlation depending on the distance between sites.

6.15 If the data were analyzed using the difference in PCB content (1996 data - 1982 data) at each site, the effect of between site variability could potentially be reduced. The data should be analyzed as a *before and after* study using paired data methodology.

6.17   a. $H_o : \mu_M \geq \mu_C$ versus $H_a : \mu_M < \mu_C$;

$$t = \frac{12690 - 6458}{\sqrt{\frac{(890)^2}{200} + \frac{(250)^2}{200}}} = 95.34 \Rightarrow \quad \text{with } df \approx 230, \quad p - value < 0.0005 \Rightarrow$$

reject H_o and conclude the data provide sufficient evidence that the California has a higher mean cost.

    b. We are 95% confident that the difference in mean cost is between $6104 and $6360.

    c. The separate-variance t-test was used since s_C^2 is 12.7 times larger than s_M^2.

6.3 A Nonparametric Alternative: The Wilcoxon Rank Sum Test

6.19   a. Plumber 1: $\mu_1 = 88.81, s_1 = 7.89$

         Plumber 2: $\mu_2 = 108.93, s_2 = 8.73$

   b. The box plots are given here:

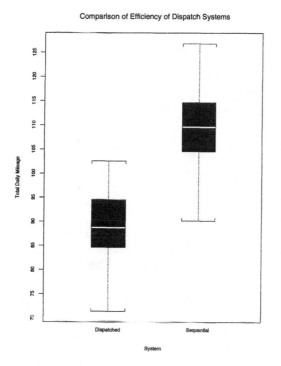

Comparison of Efficiency of Dispatch Systems

Since both graphs show a roughly symmetrical distribution with no outliers, a t-test appears to be appropriate.

6.21   a. Box plots are given here:

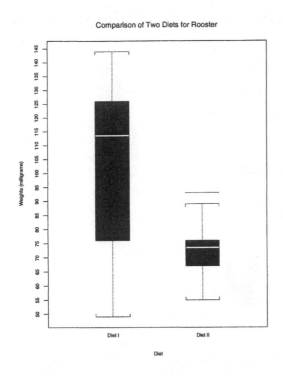

Comparison of Two Diets for Rooster

Both data sets appear to be random samples from normally distributed populations, although there is an outlier in the Diet II data set. However, the variance from Diet I is nearly 8 times larger than the variance from Diet II. Thus, the pooled t test would not be appropriate.

b. $t = \frac{105.7 - 73.1}{\sqrt{\frac{(28.3)^2}{14} + \frac{(10.1)^2}{14}}} = 4.07 \Rightarrow$ with $df \approx 16$, $0.0005 < p - value < 0.001 \Rightarrow$

Reject H_o and conclude the data provide sufficient evidence that there is a significant difference in average comb weights.

c. Using the Wilcoxon rank sum test with $\alpha = 0.05$, reject H_o if $|z| \geq 1.96$, where $z = (T - \mu_T)/\sigma_T$.

$T = 266.5, \mu_T = 14(14 + 14 + 1)/2 = 203, \sigma_T^2 = (14)(14)(14 + 14 + 1)/12 = 473.67,$

$z = (266.5 - 203)/\sqrt{473.67} = 2.92 \Rightarrow p - value = 2P(|z| \geq 2.92) = 0.0035.$

Reject H_o and conclude there is significant evidence of a difference in average comb weights from the two diets.

d. Because the variances differ greatly, the Wilcoxon rank sum test is inappropriate. The distributions both appear to be normally distributed, thus the separate variance t test is the most appropriate test statistic.

6.22   a. The data from the treatment group appears to be a random sample from a normal distribution but the data from the control group is definitely not normally distributed. The variance from the control group is 13.4 times larger than the variance from the treatment group. The Wilcoxon rank sum test is probably the most appropriate test

48

statistic since it is more robust to deviations from stated conditions than is the separate variance t test.

b. The p-value from the separate variance t test is 0.053 and the p-value from the Wilcoxon test is 0.0438. With $\alpha = 0.05$, the conclusions from the two tests differ with the t test failing to find that the mean daily mileage for the treatment group significantly smaller than the mean for the control group. The Wilcoxon did find a significant reduction in the mean of the treatment group.

c. The t test is probably not appropriate since the control group appears to have a nonnormal distribution.

d. The choice of test statistic would definitely have an effect on the final conclusions of the study. When the test statistics yield contradictory results, and the required conditions of the test statistics are not met, it is best to error on the side of the most conservative conclusion relative to which of the two Types of errors has the greatest consequence. Hence, reject H_o only if one of the tests is highly significant relative to specified α.

6.4 Inferences about $\mu_1 - \mu_2$: Paired Data

6.27 a. To conduct the study using independent samples, the 30 participants should be very similar relative to age, body fat percentage, diet, and general health prior to the beginning of the study. The 30 participants would then be randomly assigned to the two treatments.

b. The participants should be matched to the greatest extend possible based on age, body fat, diet, and general health before the treatment is applied. Once the 15 pairs are configured, the two treatments are randomly assigned within each pair of participants.

c. If there is a large difference in the participants with respect to age, body fat, diet and general health and if the pairing results in a strong positive correlation in the responses from paired participants, then the paired procedure would be more effective. If the participants are quite similar in the desired characteristics prior to the beginning of the study, then the independent samples procedure would yield a test statistic having twice as many df as the paired procedure and hence would be more powerful.

6.29 a. The paired t-test yields $t = \frac{2.58}{9.49/\sqrt{10}} = 0.86, df = 9, \Rightarrow$

$p - value = P(t \geq 0.86) \Rightarrow 0.10 < p - value < 0.25$

There is not significant evidence that the mean SENS value decreased.

b. The 95% C.I. estimate of the change in the mean SENS value is (-4.21, 9.37).

c. The box plot indicates that results from patient number 9 are an outlier relative to the other patients. The remaining values appear to have a normal distribution but the results from patient number 9 should be carefully checked. The researchers should be interviewed to confirm that the results from the 9 patients are truly independent, i.e., the differences form a random sample from a normal distribution.

6.5 A Nonparametric Alternative: The Wilcoxon Signed-Rank Test

6.35 a. The box plot and normal probability plots both indicate that the distribution of the data is somewhat skewed to the left. Hence, the Wilcoxon would be more appropriate, although the paired t-test would not be inappropriate since the differences are nearly normal in distribution.

  b. H_o : The distribution of differences (female minus male) is symmetric about 0 versus
  H_a : The differences (female minus male) tend to be larger than 0
  With $n = 20, \alpha = 0.05, T = T_-$, reject H_o if $T_- \leq 60$.
  From the data we obtain $T_- = 18 < 60$, thus reject H_o and conclude that repair costs are generally higher for female customers.

Supplementary Exercises

6.37 a. Population of interest is the collection of all steel beams made from both the new and old alloy.

  b. 99% C.I. on Old Alloy Mean Load Capacity: (22.12, 24.62)
  99% C.I. on New Alloy Mean Load Capacity: (26.27, 31.43)
  99% C.I. on $\mu_{Old} - \mu_{New}$: (-8.14, -2.82)

  c. $H_o : \mu_{Old} \geq \mu_{New}$ versus $H_a : \mu_{Old} < \mu_{New}$;
  $t = \dfrac{28.85 - 23.37}{\sqrt{\frac{(2.51)^2}{10} + \frac{(1.22)^2}{10}}} = 6.21 \Rightarrow$ with $df \approx 13$, $p-value < 0.0005 \Rightarrow$

  reject H_o and conclude the data provide sufficient evidence that the new load has a higher mean load capacity.

  d. Box plots are given here:

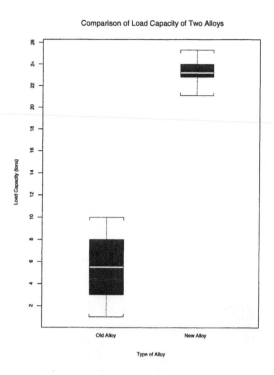

Comparison of Load Capacity of Two Alloys

The separate-variance t-test was used since s^2_{New} is 4.2 times larger than s^2_{Old}, but the distributions both appear normally distributed.

   e. The company should not switch alloys because the C.I. on $\mu_{Old} - \mu_{New}$ indicates that the difference may be as small as -2.82 units.

6.39   a. $H_o : \mu_{Narrow} = \mu_{Wide}$ versus $H_a : \mu_{Narrow} \neq \mu_{Wide}$;

$$t = \frac{118.37 - 110.20}{\sqrt{\frac{(7.87)^2}{12} + \frac{(4.71)^2}{15}}} = 3.17 \Rightarrow \quad \text{with} \quad df \approx 17, \quad 0.005 < p-value < 0.01 \Rightarrow$$

reject H_o and conclude there is sufficient evidence in the data that the two types of jets have different average noise levels.

   b. A 95% C.I. on $\mu_{Wide} - \mu_{Narrow}$ is (2.73, 13.60)

   c. Because maintenance could affect noise levels, jets of both types from several different airlines and manufacturers should be selected. They should be of approximately the same age, etc. This study could possibly be improved by pairing Narrow and Wide body airplances based on factors that may affect noise level.

6.42   a. $H_o : \mu_{Within} = \mu_{Out}$ versus $H_a : \mu_{Within} \neq \mu_{Out}$;

$$t = \frac{3092 - 2450}{\sqrt{\frac{(1191)^2}{14} + \frac{(2229)^2}{12}}} = 0.89 \Rightarrow \quad \text{with} \quad df \approx 16, \quad p-value \approx 0.384 \Rightarrow$$

Fail to reject H_o and conclude the data does not provide sufficient evidence that there is a difference in average population abundance.

   b. A 95% C.I. on $\mu_{Within} - \mu_{Out}$ is (-879, 2164)

c. The two samples are independently selected random samples from two normally distributed populations.

d. Box plots are given here:

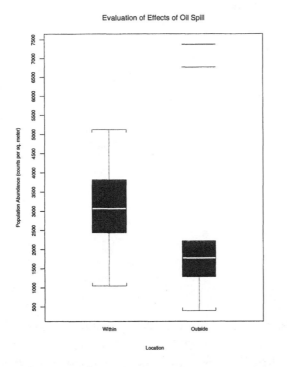

Evaluation of Effects of Oil Spill

The Within data set appears to be normally distributed but the Outside data may not be normally distributed since there were two outliers. The sample variance of the Outside is 3.5 times larger than the Within sample variance.

6.43   a. $H_o : \mu_{Within} = \mu_{Out}$ versus $H_a : \mu_{Within} \neq \mu_{Out}$;
Since both n_1, n_2 are greater than 10, the normal approximation can be used.
$T = 122, \mu_T = (12)(12 + 14 + 1)/2 = 162, \sigma = \sqrt{(12)(14)(12 + 14 + 1)/12} = 19.44$
$z = \frac{122 - 162}{19.44} = 2.06 \Rightarrow p - value = 0.0394 \Rightarrow$
reject H_o and conclude the data provides sufficient evidence that there is a difference in average population abundance.

b. The Wilcoxon rank sum test requires independently selected random samples from two populations which have the same shape but may be shifted from one another.

c. The two population distributions may have different variances but the Wilcoxon rank sum test is very robust to departures from the required conditions.

d. The separate variance test failed to reject H_o with a p-value of 0.384. The Wilcoxon test rejected H_o with a p-value of 0.0394. The difference in the two procedures is probably due to the skewness observed in the Outside data set. This can result in inflated p-values for the t test which relies on a normal distribution when the sample sizes are small.

6.44   a. Box plot is given here:

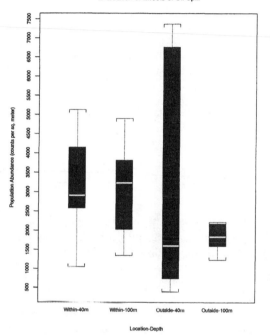

Evaluation of Effects of Oil Spill

b. 40M: $H_o : \mu_{Within} = \mu_{Out}$ versus $H_a : \mu_{Within} \neq \mu_{Out}$;

$$ t = \frac{3183 - 3080}{\sqrt{\frac{(1289)^2}{7} + \frac{(3135)^2}{6}}} - 0.08 \Rightarrow \quad \text{with} \quad df \approx 6, \quad p - value \approx 0.94 \Rightarrow $$

fail to reject H_o and conclude the data does not provide sufficient evidence that there is a difference in average population abundance at 40 m.

100M: $H_o : \mu_{Within} = \mu_{Out}$ versus $H_a : \mu_{Within} \neq \mu_{Out}$;

$$ t = \frac{3002 - 1820}{\sqrt{\frac{(1181)^2}{7} + \frac{(385)^2}{6}}} = 2.50 \Rightarrow \quad \text{with} \quad df \approx 7, \quad p - value \approx 0.041 \Rightarrow $$

reject H_o and conclude the data provides sufficient evidence that there is a difference in average population abundance at 100 m.

c. The conclusions are different at the two depths. The mean population abundances are fairly consistent at all but the 100 m depth outside the oil trajectory where the abundance is considerably smaller. However, the median abundance Within are nearly the same at both Depths; but both Within medians are higher than the median abundances at the two Outside depths.

6.45   a. Such a statement cannot be made. In order to study the effect of the oil spill on the population, we would need some baseline data at these sites before the oil spill.

b. If baseline data were available, a paired t-test would be appropriate. The impact of the oil spill could be studies by a *before and after* analysis. If the pairing were effective, the paired data study would be more efficient than a two independent samples study, i.e., it would take fewer observations (sites) to achieve the same level of precision.

c. Spatial correlation might be a problem. Factors such as weather, food supply, etc., which are outside the control of the researcher might mask or exagerate the effect of the oil spill. These types of factors also prevent the researcher from making definitive *cause and effect* statements.

6.49 a. $H_o : \mu_{D_1} = \mu_{D_2}$ versus $H_a : \mu_{D_1} \neq \mu_{D_2}$;
Since $s_1 = 13.49$ and $s_2 = 18.49$, there is not strong evidence that the population variances are unequal, hence use a pooled t-test:
$t = \frac{34.24 - 44.03}{16.184 \sqrt{\frac{1}{10} + \frac{1}{10}}} = -1.35$ with $df = 18$, $p-value \approx 0.19$. $\Rightarrow$
Fail to reject H_o and conclude the data fails to provide sufficient evidence that the two drugs have different mean absorption time.

b. p-value $= 2P(t \geq |-1.35|) \approx 0.19$

c. 95% C.I. on $\mu_{D_1} - \mu_{D_2}$: (-25.0, 5.4) minutes, i.e., With 95% confidence, drug D_1 takes between 25 minutes less to 5.4 minutes more to reach the specified blood level than does drug D_2.

d. The random samples were independently selected and the population variances are not significantly different. The box plots and normal probability plots do not indicate deviations from a normally distributed populations.

6.51 a. $H_o : \mu_{High} = \mu_{Con}$ versus $H_a : \mu_{High} \neq \mu_{Con}$;
Separate variance t-test: $t = 4.12$ with $df \approx 34$, $p-value = 0.0002$. $\Rightarrow$
Reject H_o and conclude there is significant evidence of a difference in the mean drop in blood pressure between the high-dose and control groups.

b. 95% C.I. on $\mu_{High} - \mu_{Con}$: (19.5, 57.6), i.e., the high dose groups mean drop in blood pressure was, with 95% confidence, 19.5 to 57.6 points greater than the mean drop observed in the control group.

c. Provided the researcher independently selected the two random samples of participants, the conditions for using a separate variance t-test were satisfied since the plots do not detect a departure from a normally distribution but the sample variances are somewhat different(1.9 to 1 ratio).

6.52 a. $H_o : \mu_{Low} = \mu_{Con}$ versus $H_a : \mu_{Low} \neq \mu_{Con}$;
Separate variance t-test: $t = -2.09$ with $df \approx 35$, $p-value = 0.044$. $\Rightarrow$
Reject H_o and conclude there is significant evidence of a difference in the mean drop in blood pressure between the low-dose and control groups.

b. 95% C.I. on $\mu_{Low} - \mu_{Con}$: (-51.3, -0.8), i.e., the low-dose groups mean drop in blood pressure was, with 95% confidence, 51.3 to 0.8 points less than the mean drop observed in the control group.

c. Provided the researcher independently selected the two random samples of participants, the conditions for using a pooled t-test were satisfied since the plots do not detect a departure from a normally distribution but the sample variances are somewhat different(1.7 to 1 ratio).

6.53   a. $H_o : \mu_{Low} = \mu_{High}$ versus $H_a : \mu_{Low} \neq \mu_{High}$;

Separate variance t-test: $t = -5.73$ with $df \approx 29$, $p-value < 0.0005$. $\Rightarrow$

Reject H_o and conclude there is significant evidence of a difference in the mean drop in blood pressure between the high-dose and low-dose groups.

b. 95% C.I. on $\mu_{Low} - \mu_{High}$: (-87.6, -41.5), i.e., the low-dose group mean drop in blood pressure was, with 95% confidence, 41.5 to 87.6 points lower than the mean drop observed in the high-dose group.

c. Provided the researcher independently selected the two random samples of participants, the conditions for using a pooled t-test were satisfied since the plots do not detect a departure from a normally distribution and the sample variances are somewhat different(3.2 to 1 ratio).

6.54   a. Let A_i be the event that a Type I error was made on the *ith* test, $i = 1, 2, 3$.

P(at least one Type I error in 3 tests) $= P(A_1 \cup A_2 \cup A_3)$

$= P(A_1) + P(A_2) + P(A_3) - P(A_1 \cap A_2) - P(A_1 \cap A_3) - P(A_2 \cap A_3) + P(A_1 \cap A_2 \cap A_3)$

$\leq P(A_1) + P(A_2) + P(A_3) = 0.05 + 0.05 + 0.05 = 3(0.05) = 0.15$

b. Set the value of α at $\frac{0.05}{3}$ for each of the 3 tests.

Then P(at least one Type I error in 3 tests) $\leq 3(\frac{0.05}{3}) = 0.05$.

Thus, we know that the chance of at least one Type I error in the 3 tests is at most 0.05.

6.59   a. $H_o : \mu_D \leq \mu_{RN}$ versus $H_a : \mu_D > \mu_{RN}$;

Separate variance t-test: $t = 9.04$ with $df \approx 76$, $p-value < 0.0001 \Rightarrow$

Reject H_o and conclude there is significant evidence that the mean score of the Degreed nurses is higher than the mean scores of the RN nurses.

b. $p-value < 0.0001$

c. 95% C.I. on $\mu_D - \mu_{RN}$: (35.3, 55.2) points

d. Since 40 is contained in the 95% C.I., there is a possibility that the difference in mean scores is less than 40. Hence, the differences observed may not be meaningful.

6.63   a. $H_o : \mu_F \geq \mu_M$ versus $H_a : \mu_F < \mu_M$;

b. 95% C.I. on $\mu_F - \mu_M$: (-142.30, -69.1) thousands of dollars

c. Since $s_1 \approx s_2$, use pooled t-test: $t = \dfrac{245.3 - 350.1}{57.2\sqrt{\frac{1}{20} + \frac{1}{20}}} = -5.85$ with $df = 38$

$p-value < 0.0001 \Rightarrow$

Reject H_o and conclude there is significant evidence that the mean female campaign expenditures are less than mean male candidates expenditures.

d. Yes, since the difference could be as much as $142,300.

6.64 The required conditions are that the two samples are independently selected from populations having normal distributions with equal variances. The box plots do not reveal any indication that the population distributions were not normal. The sample variances have a ratio of 1.4 to 1.0, thus there is very little indication that the population variances were unequal.

6.67   a. $H_o : \mu_{Standard} \geq \mu_{New}$ versus $H_a : \mu_{Standard} < \mu_{New}$

Since $s_1 \approx s_2$, use pooled t-test: $t = -1.94$ with $df = 54$, $p-value = 0.029 \Rightarrow$ Reject H_o and conclude there is significant evidence that the new therapy has a higher mean survival time.

b. 95% C.I. on $\mu_{New} - \mu_{Standard}$: $(0.2, 10.2)$

c. The required conditions are that the two samples are independently selected from populations having normal distributions with equal variances. The box plots are fairly consistent that the population distributions were are normal, although the distribution of the survival times for the new therapy is slightly right skewed. The sample variances have a ratio of 1.04 to 1.0, thus there is very little indication that the population variances were unequal.

6.69   a. $H_o : \mu_{Last} \geq \mu_{Current}$ versus $H_a : \mu_{Last} < \mu_{Current}$

b. $t = \dfrac{320-410}{\sqrt{\frac{300}{100} + \frac{350}{100}}} = -35.3$, With df $= 198 \Rightarrow p-value \approx 0$

Reject H_o and conclude there is sufficient evidence that the mean refund has increased from the previous year.

c. Since the sample sizes are quite large, the only condition that must be satisfied is that the two samples are independently selected random samples.

6.71 $H_o : \mu_{For} \leq \mu_{USA}$ versus $H_a : \mu_{For} > \mu_{USA}$

Since $s_1 \approx s_2$, use separate-variance t-test: $t = \dfrac{207.45 - 182.54}{39.4286\sqrt{\frac{1}{14} + \frac{1}{14}}} = 1.67$, With df $= 26 \Rightarrow$

$0.05 < p-value < 0.10$

Fail to reject H_o and conclude the data does not support the claim that foreign-produced selections have longer mean playing times.

6.73   a. The average potency after 1 year is different than the average potency right after production.

b. The two test statistics are equal since the sample sizes are equal.

c. The p-values are different since the test statistics have different degrees of freedom (df).

d. In this particular experiment, the test statistics reach the same conclusion, reject H_o.

e. Since $s_1 \approx s_2$ which yields a test of equal variances with p-value of 0.3917, the pooled t-test would be the more appropriate test statistic.

6.77   a. A box plot is given here:

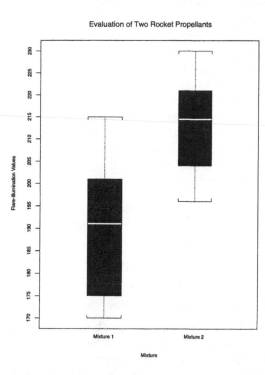

Evaluation of Two Rocket Propellants

Since the box plots indicate normally distributed data and $s_1 = 14.652$ and $s_2 = 10.627$ are not very different, use pooled t-test.

b. $t = \frac{190 - 213.4}{12.64\sqrt{\frac{1}{10} + \frac{1}{10}}} = -4.09$, With df $= 18 \Rightarrow p - value < 0.0005$

Reject H_o and conclude there is significant evidence that the mean flare-illumination values is different for the two mixtures.

6.78 A 95% C.I. on the difference in the mean flare-illumination values,

$$\mu_1 - \mu_2 : (190 - 213.4) \pm (2.101)(12.64)\sqrt{\frac{1}{10} + \frac{1}{10}} \Rightarrow (-35.425, -11.375)$$

6.79 $n = \frac{(12)^2(1.645 + 1.282)^2}{(15)^2} = 5.5 \Rightarrow n = 6$ for each mixture.

6.83 d = Analyst 2 - Analyst 1

a. $H_o : \mu_d \geq 0$ versus $H_a : \mu_d < 0$

use paired t-test:$t = \frac{-1.383}{1.853/\sqrt{6}} = -1.83$, With df $= 5 \Rightarrow 0.05 < p - value < 0.10$

Fail to reject H_o and conclude there is not significant evidence that analyst 1 reads higher on the average than analyst 2.

b. Wilcoxon signed-rank test: Reject H_o using $\alpha = 0.05$ if $T_+ < 2$. From the data $T_+ = 3$, thus fail to reject H_o. This result agrees with our findings using the paired t-test.

6.89 a. Baseline findings indicate that nonsurvivors have more severe heart failures than survivors do. For each measure of heart failure, the difference between the two groups is significant at least a 0.01 level with survivors having the less severe measurement. No significant differences occur in measuring age, duration of symptoms or heart rate.

b. In making these t-tests, the authors assume that the samples are independently selected random samples from populations of survivors and nonsurvivors which are approximately normally distributed with the population variances nearly equal. The last assumption is not too restrictive here as the sample sizes are nearly equal, thus allowing the variances to differ somewhat without invalidating the conclusions. Because there was no randomization of the groups and the study compares survivors and nonsurvivors, we must be careful about inferring a casual relationship. Other confounding factors may have been neglected by this procedure; that is, the two groups may differ in some unnoticed way that causes one to survive and the other to die.

6.93 The results as stated in the problem indicate nothing statistically. Although we are given the sample size, we are given neither the average decreases nor their standard errors. Thus we can make no conclusion about the significance of the results.

6.94 a. Reporting a standard error would provide some indication of the variability in the sample and thus allow some comparison of the means of the two groups.

b. A t-test for a comparison of two means is the most likely test used in this situation.

c. The data here contain no tests of arithmetic or reading ability. The IQ test measures overall ability but does not relate specifically to reading or arithmetic. Some indication of the type of foreign language spoken could affect scores. For example, those from families speaking languages close to English like German or French will have less difficulty than those speaking Chinese or Japanese. Another characteristic which may influence scores is the length of residence in the country. Specifically, one could determine which children were born in Australia.

d. Students who leave before the sixth year of primary school either drop out or move away from the study area. Those who drop out are more likely to be from poorer families who do not value or cannot afford to send their child to school. In either case, such students probably perform worse than those who remain in school. Children whose families move may come from migrant families or from professional ones whose parents are transferred. Students from migrant families probably do worse. Attrition probably increases scores on the whole, especially among the non-English speaking families.

Chapter 7: Inferences about Population Variances

7.2 Estimation and Tests for a Population Variance

7.1   a. 0.01

     b. 0.90

     c. 1 - 0.99 = 0.01

     d. 1 - 0.01 - 0.01 = 0.98

7.3   a. $\chi^2_{.025} \approx 80 \left(1 - \frac{2}{(9)(80)} + 1.96\sqrt{\frac{2}{(9)(80)}}\right)^3 = 106.632$

       $\chi^2_{.975} \approx 80 \left(1 - \frac{2}{(9)(80)} - 1.96\sqrt{\frac{2}{(9)(80)}}\right)^3 = 57.1462$

       Rounded to 2 decimal places, the above approximations are exactly the values given in Table 7 for the Chi-square distribution.

     b. $\chi^2_{.025} \approx 277 \left(1 - \frac{2}{(9)(277)} + 1.96\sqrt{\frac{2}{(9)(277)}}\right)^3 = 324.999$

       $\chi^2_{.975} \approx 277 \left(1 - \frac{2}{(9)(277)} - 1.96\sqrt{\frac{2}{(9)(277)}}\right)^3 = 232.787$

7.5   a. Let y be the quantity in a randomly selected jar:

       Proportion $= P(y < 32) = P(z < \frac{32-32.3}{.15}) = 0.0228 \Rightarrow 2.28\%$

     b. The plot indicates that the distribution is approximately normal because the data values are reasonably close to the straight-line.

     c. 95% C.I. on σ : $\left(\sqrt{\frac{(50-1)(.135)^2}{70.22}}, \sqrt{\frac{(50-1)(.135)^2}{31.55}}\right) \Rightarrow (0.113, 0.168)$

     d. $H_o : \sigma \leq 0.15$ versus $H_a : \sigma > 0.15$

       Reject H_o if $\frac{(n-1)(s)^2}{(.15)^2} \geq 66.34$

       $\frac{(50-1)(.135)^2}{(.15)^2} = 39.69 < 66.34 \Rightarrow$

       Fail to reject H_o and conclude the data does not support σ greater than 0.15.

     e. p-value $= P(\frac{(n-1)(s)^2}{(.15)^2} \geq 39.69)$

       Using the Chi-square tables with $df = 49, 0.10 < p - value < 0.90$

       (Using a computer program, p-value $= 0.8262$).

7.7   a. $n = 26, \quad \bar{y} = 3.999, \quad s = 0.0159$

     b. $H_o : \sigma \leq 0.011$ versus $H_a : \sigma > 0.011$

       Reject H_o if $\frac{(n-1)(s)^2}{(.011)^2} \geq 37.65$

       $\frac{(26-1)(.0159)^2}{(.011)^2} = 52.23 > 37.65 \Rightarrow$

       Reject H_o and conclude the data supports the statement that σ greater than 0.011.

7.8 90% C.I. on σ : $\left(\sqrt{\frac{(26-1)(0.0159)^2}{37.65}}, \sqrt{\frac{(26-1)(0.0159)^2}{14.61}}\right) \Rightarrow (0.01297, 0.02080)$

7.9 The box plot and normal probability plot is given here:

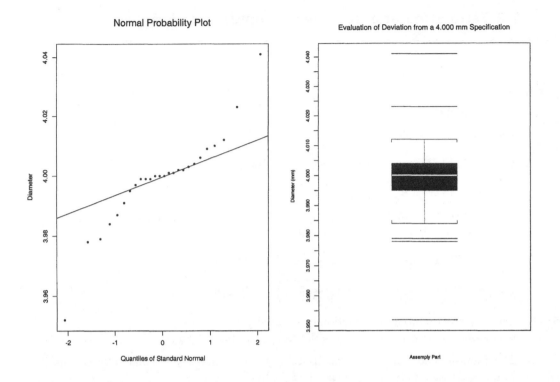

The box plot is symmetric but there are four outliers. This would indicate that the population distribution may have heavier tails than a normal distribution. The normal probability plot has values deviating from the straight-line. This may cause the values of s to be inflated. Therefore, the level of significance may be inflated and the C.I.'s may be too wide for the given level of confidence.

7.3 Estimation and Tests for Comparing Two Population Variances

---

7.13 a. $\alpha = .05, df_1 = 7, df_2 = 12, \Rightarrow F_{0.05} = 2.91$

  b. $\alpha = .05, df_1 = 3, df_2 = 10, \Rightarrow F_{0.05} = 3.71$

  c. $\alpha = .05, df_1 = 10, df_2 = 20, \Rightarrow F_{0.05} = 2.35$

  d. $\alpha = .01, df_1 = 8, df_2 = 15, \Rightarrow F_{0.01} = 4.00$

  e. $\alpha = .01, df_1 = 12, df_2 = 25, \Rightarrow F_{0.01} = 2.99$

7.15 We need to assume that the two samples were independently selected from normally distributed populations.

$H_o : \sigma_1^2 = \sigma_2^2$ versus $H_a : \sigma_1^2 \neq \sigma_2^2$

With $\alpha = 0.10$, reject H_o if $\frac{s_1^2}{s_2^2} \leq \frac{1}{3.68} = 0.272$ or $\frac{s_1^2}{s_2^2} \geq 3.29$

$s_1^2/s_2^2 = 0.583 \Rightarrow 0.272 < 0.583 < 3.29 \Rightarrow$

Fail to reject H_o and conclude the data does not support the contention that the population variances are different.

7.17   a. 95% C.I. on σ_{Old} : $\left(\sqrt{\frac{(61-1)(0.231)^2}{83.30}}, \sqrt{\frac{(61-1)(0.231)^2}{40.48}} \right) \Rightarrow (0.196, 0.281)$

     95% C.I. on σ_{New} : $\left(\sqrt{\frac{(61-1)(0.162)^2}{83.30}}, \sqrt{\frac{(61-1)(0.162)^2}{40.48}} \right) \Rightarrow (0.137, 0.197)$

  b. $H_o : \sigma_{Old}^2 \leq \sigma_{New}^2$ versus $H_a : \sigma_{Old}^2 > \sigma_{New}^2$

    With $\alpha = 0.05$, reject H_o if $\frac{s_{Old}^2}{s_{New}^2} \geq 1.53$

    $s_{Old}^2/s_{New}^2 = 2.033 > 1.53 \Rightarrow$

    Reject H_o and conclude the data supports the statement that σ_{New}^2 is less than σ_{Old}^2.

  c. The box plots indicate the both population distributions are normally distributed. From the problem description, the two samples appear to be independently selected random samples.

7.4 Tests for Comparing $t > 2$ Population Variances

7.21   a. From the box plots, the shapes are symmetric with equal length whiskers and no outliers. Thus, we would conclude that the data is from normally distributed populations.

  b. With $\alpha = 0.01$, reject H_o if $F_{max} > 9.9$

    $F_{max} = \frac{80.22}{6.89} = 11.64 > 9.9 \Rightarrow$

    Reject H_o at level $\alpha = 0.01$ and conclude there is a significant difference in the population variances.

    The Levine test yields $L = 4.545$ with $df_1 = (3-1) = 2, df_2 = (27-3) = 24$

    With $\alpha = 0.01$, reject H_o if $L > F_{.01,2,24} = 5.61$

    $L = 4.545 < 5.61 \Rightarrow$ Fail to reject H_o at the 0.01 level.

    Thus, the Levine and Hartley test yield contradictory conclusions.

  c. When the population distributions are normally distributed, the Hartley test is more powerful than the Levine test. Thus, in this situation, we would prefer the Hartley test.

  d. 95% C.I. on σ_1 : $\left(\sqrt{\frac{(9-1)(8.69)}{17.53}}, \sqrt{\frac{(9-1)(8.69)}{2.18}} \right) \Rightarrow (1.99, 5.65)$

     95% C.I. on σ_2 : $\left(\sqrt{\frac{(9-1)(6.89)}{17.53}}, \sqrt{\frac{(9-1)(6.89)}{2.18}} \right) \Rightarrow (1.77, 5.03)$

95% C.I. on σ_3: $\left(\sqrt{\frac{(9-1)(80.22)}{17.53}}, \sqrt{\frac{(9-1)(80.22)}{2.18}}\right) \Rightarrow (6.05, 17.16)$

Based on the C.I.'s and the fact that we concluded there was a significant difference in the variances for the three additives, we can conclude that Additives 1 and 2 yield similar levels of precision because their C.I.'s overlap considerably. However, Additive 3 yields considerably higher levels of variability which would probably exclude it from use in the process.

7.25 The data is summarized in the following table:

Method	n	Mean	95% C.I. on μ	St.Dev.	95% C.I. on σ
L	10	5.90	(2.34, 9.46)	4.9766	(3.42, 9.09)
L/R	7	7.29	(2.31, 12.26)	5.3763	(3.46, 11.84)
L/C	9	16.00	(9.57, 22.43)	8.3666	(5.65, 16.03)
C	9	17.67	(5.45, 29.88)	15.8902	(10.73, 30.44)

From the box plots, it appears that the L/R distribution is right skewed but the other 3 distributions appear to be random samples from normal distributions.

Test $H_o : \sigma_L = \sigma_{L/R} = \sigma_{L/C} = \sigma_C$ versus $H_a : \sigma's$ are different

Reject H_o at level $\alpha = 0.05$ if $L \geq F_{.05,3,31} = 2.91$

From the data, $L = 2.345 < 2.91 \Rightarrow$

Fail to reject H_o and conclude there is not significant evidence of difference in variability of the increase in test scores.

Based on the C.I.'s for the $\mu's$, we can conclude that there is very little difference in the average change in test scores for the four methods of instruction. However, lecture only method yielded somewhat smaller mean change in test score than the computer instruction only procedure. These confidence intervals have an overall level of confidence of $(.95)^4 = 0.81$ since the data from the four procedures are independent. Thus, our conclusion would have a relatively large chance of committing a Type I error in attempting to determine if any pair of instructional methods have different means. An improved procedure for comparing the four instructional methods will be covered in Chapter 8. This procedure would determine that there is a significant difference in the instructional means (p-value = 0.032).

7.27 a. The box plots are given here:

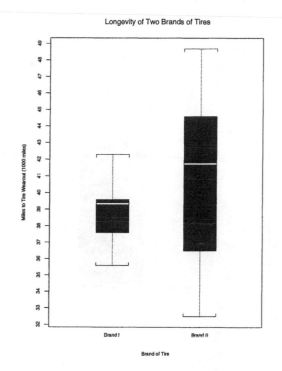

Longevity of Two Brands of Tires

The box plots and normal probability plots indicate that both samples are from normally distributed populations.

b. The C.I.'s are given here:

Method	n	Mean	95% C.I. on μ	St.Dev.	95% C.I. on σ
I	10	38.79	(37.39, 40.19)	1.9542	(1.34, 3.57)
II	10	40.67	(36.68, 44.66)	5.5791	(3.84, 10.19)

c. A comparison of the population variances yields:

$H_o : \sigma_I^2 = \sigma_{II}^2$ versus $H_a : \sigma_I^2 \neq \sigma_{II}^2$

With $\alpha = 0.01$, reject H_o if $\frac{s_I^2}{s_{II}^2} \leq \frac{1}{6.54} = 0.15$ or $\frac{s_1^2}{s_2^2} \geq 6.54$

$s_1^2/s_2^2 = (5.5791)^2/(1.9542)^2 = 8.15 > 6.54 \Rightarrow$

Reject H_o and conclude there is significant evidence that the population variances are different.

A comparison of the population means using the separate variance t-test yields:

$H_o : \mu_I = \mu_{II}$ versus $H_a : \mu_I \neq \mu_{II}$

$t = \dfrac{38.79 - 40.67}{\sqrt{\frac{(1.9542)^2}{10} + \frac{(5.5791)^2}{10}}} = -1.01$ with df=11 $\Rightarrow p - value = 0.336$

Fail to reject H_o and conclude that the data does not support a difference in the mean tread wear for the two brands of tires. However, Brand I has a more uniform tread wear as reflected by it significantly lower standard deviation.

7.29   a. 25x90% = 22.5 and 25x110% = 27.5 implies the limits are 22.5 to 27.5

b. The box plot and normal probability plot are given here:

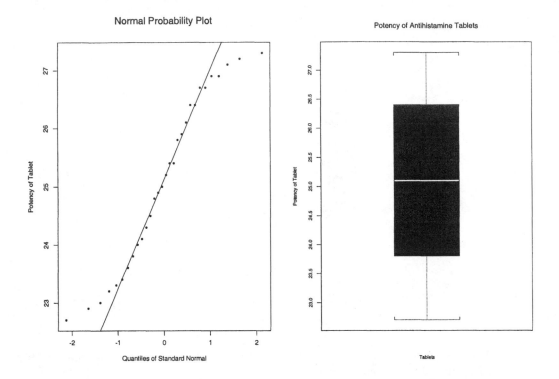

The box plot indicates a symmetric distribution with no outliers. The normal probability plot shows the data values reasonably close to a straight line, although there is some deviation at both ends which indicates that the data may be a random sample from a distribution which has shorter tails than a normally distributed population.

c. Range = 27.5-22.5 = 5 $\Rightarrow \hat{\sigma} = 5/4 = 1.25$

$H_o : \sigma = 1.25$ versus $H_a : \sigma \neq 1.25$

With $\alpha = 0.05$, reject H_o if $\frac{(n-1)s^2}{(1.25)^2} \leq 16.05$ or $\frac{(n-1)s^2}{(1.25)^2} \geq 45.72$

$\frac{(30-1)(1.4691)^2}{(1.25)^2} = 40.06 \Rightarrow 16.06 < 40.06 < 45.72$

Fail to reject H_o and conclude there is insufficient evidence that the product standard deviation is greater than 1.25. Thus, it appears that the potencies are within the required bounds.

7.31 95% C.I. on σ^2 : $\left(\frac{(20-1)5.45}{32.852}, \frac{(20-1)5.45}{8.907} \right) \Rightarrow (3.15, 11.63)$

The 95% C.I. fails to support the test findings of the consumer group since the C.I. contains values less than 4 and the consumer group was attempting to determine whether or not σ^2 was greater than 4. The test of Example 7.2 does not have much power to detect an increase in σ^2 of 25% over the claimed value because the sample size was relatively small which leads to a fairly wide C.I.. The C.I. and test yield contradictory results since the test was 1-tail test whereas the C.I. is equivalent to a 2-tail test.

7.33 Since the box plots indicate that data from both portfolios has a normal distribution. Also, the C.I. on the ratio of the variances contained 1 which indicates equal variances. Thus, a pooled variance t-test will be used as the test statistic.

$H_o : \mu_1 = \mu_2$ versus $H_a : \mu_1 \neq \mu_2$

$t = \frac{131.60 - 147.20}{4.92\sqrt{\frac{1}{10} + \frac{1}{10}}} = -7.09$ with df=18 $\Rightarrow p - value < 0.0005$

Reject H_o and conclude that the data strongly supports a difference in the mean returns of the two portfolios.

7.37 The box plots and normal probability plots indicate that both samples are probably not from normally distributed populations.

Summary statistics are given here: (C.I.'s are given for the median. C.I.'s for $\mu's$ and $\sigma's$ are not appropriate because the distributions are nonnormal and the sample sizes are relatively small)

Method	n	Mean	Median	95% C.I. on Median	St.Dev.
EXP	15	60.20	55.8	(51.4, 69.0)	10.8407
MAR	15	53.63	52.7	(50.1, 55.8)	3.9808

The Wilcoxon test of differences in the two distributions yields

$T = 280.5, \mu_T = 15(15 + 15 + 1)/2 = 232.5, \sigma_T = \sqrt{(15)(15)(15 + 15 + 1)/12} = 24.109 \Rightarrow$

$z = \frac{280.5 - 232.5}{24.109} = 1.99 \Rightarrow p - value = 2P(z \geq 1.99) = 0.0465 < 0.05 \Rightarrow$

The data indicates that the two drugs have different distributions for the potency drop.

Box plots are given here:

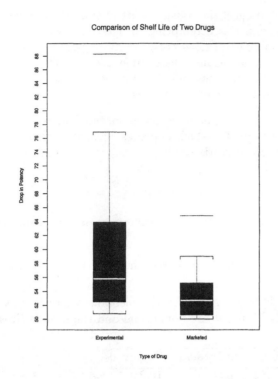

From the box plots and the values of the sample standard deviations, the experimental drug has a more widely dispersed potency drop than the marketed drug.

7.39 Box plots are given here:

Comparison of Weight-Reducing Agents

The box plots indicate that both samples are from normally distributed populations but with different levels of variability.

C.I.'s are given here:

Method	n	Mean	95% C.I. on μ	St.Dev.
A	13	27.62	(31.68, 33.55)	9.83
B	13	34.69	(32.26, 37.13)	4.03

A comparison of the population variances yields:

$H_o : \sigma_A^2 = \sigma_B^2$ versus $H_a : \sigma_A^2 \neq \sigma_B^2$

$s_A^2 / s_B^2 = (9.83)^2 / (4.03)^2 = 5.955 \Rightarrow .002 < p - value < .010 \Rightarrow$

Reject H_o and conclude there is significant evidence that the population variances are different.

A comparison of the population means using the separate variance t-test yields:

$H_o : \mu_A = \mu_B$ versus $H_a : \mu_A \neq \mu_B$

$t = \frac{27.62 - 34.69}{\sqrt{\frac{(9.83)^2}{13} + \frac{(4.03)^2}{13}}} = -2.40$ with df=15 $\Rightarrow p - value = 0.030$

Reject H_o and conclude that the data indicates a difference in the mean length of time people remain on the two therapies.

7.41 Box plots are given here:

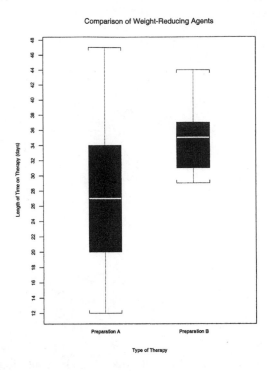

The box plots indicate that the two samples are random samples from normally distributed populations.

$H_o : \sigma_1^2 \leq \sigma_2^2$ versus $H_a : \sigma_1^2 > \sigma_2^2$

With $\alpha = 0.05$, reject H_o if $\frac{s_1^2}{s_2^2} \geq 3.18$

$s_1^2/s_2^2 = (0.7528)^2/(0.3957)^2 = 3.62 > 3.18 \Rightarrow$

Reject H_o and conclude there is significant evidence that Location 1 has a larger variance than Location 2.

95% C.I. on $\frac{\sigma_1^2}{\sigma_2^2}$: $\left(\frac{(0.7528)^2}{(0.3957)^2}(0.248), \frac{(0.7528)^2}{(0.3957)^2}(4.03) \right) \Rightarrow (0.90, 14.59)$

7.43 Box plots are given here:

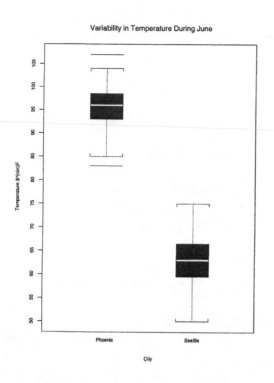

Variability in Temperature During June

A comparison of the population variances yields:

$H_o : \sigma_P^2 = \sigma_S^2$ versus $H_a : \sigma_P^2 \neq \sigma_S^2$

$s_P^2/s_S^2 = (6.4449)^2/(5.7425)^2 = 1.26 \Rightarrow p-value > .25 \Rightarrow$

Fail to reject H_o and conclude there is not significant evidence that the population variances are different.

A comparison of the population means using the pooled variance t-test yields:

$H_o : \mu_P = \mu_S$ versus $H_a : \mu_P \neq \mu_S$

$t = \frac{95.20 - 62.85}{6.10\sqrt{\frac{1}{20} + \frac{1}{20}}} = 16.76$ with df=38 $\Rightarrow p - value < 0.0005$

Reject H_o and conclude that the data indicates a difference in the mean temperature during June in the cities of Phoenix and Seattle (Very surprising?).

Chapter 8: Inferences about More Than Two Population Central Values

8.2 An Analysis of Variance

8.1   a. Yes, the mean for Device A is considerably (relative to the standard deviations) smaller than the mean for Device D.

     b. $H_o : \mu_A = \mu_B = \mu_C = \mu_D$ versus H_a : Difference in $\mu's$

       Reject H_o if $F \geq F_{.05,3,20} = 3.10$

       $SSW = 5[(.1767)^2 + (.2091)^2 + (.1532)^2 + (.2492)^2] = 0.8026$

       $\bar{y}_{..} = 0.0826 \Rightarrow$

       $SSB = 6[(-0.1605 - .0826)^2 + (0.0947 - .0826)^2 + (0.1227 - .0826)^2 + (0.2735 - .0826)^2]$

          $= 0.5838 \Rightarrow$

       $F = \frac{.5838/3}{.8026/20} = 4.85 > 3.10 \Rightarrow$

       Reject H_o and conclude there is significant difference among the mean difference in pH readings for the four devices.

     c. $p - value = P(F_{3,20} \geq 4.85) \Rightarrow 0.01 < p - value < 0.025$

     d. The data must be independently selected random samples from normal populations having the same value for σ.

     e. Suppose the devices are more accurate at higher levels of pH in the soil, and if by chance all soil samples with high levels of pH are assigned to a particular device, then that device may be evaluated as more accurate based just on the chance selection of soil samples and not on a true comparison with the other devices.

8.5 An Alternative Analysis: Transformation of the Data

8.3 $H_o : \mu_{NE} = \mu_{SE} = \mu_{NW} = \mu_{SW}$ versus H_a : Difference in $\mu's$

     Reject H_o if $F \geq F_{.05,3,20} = 3.10$

     $SSW = 5[(.2515)^2 + (.2799)^2 + (.2307)^2 + (.2693)^2] = 1.3367$

     $\bar{y}_{..} = 1.28925 \Rightarrow$

     $SSB = 6[(0.845 - 1.28925)^2 + (1.480 - 1.28925)^2 + (0.822 - 1.28925)^2 + (2.010 - 1.28925)^2]$

        $= 5.8312$

     $F = \frac{5.8312/3}{1.3367/20} = 29.08 > 3.10 \Rightarrow$

     Reject H_o and conclude there is significant difference among the mean opinions in the four geographical regions.

8.5   a. $H_o : \mu_A = \mu_B = \mu_C$ versus H_a : Difference in $\mu's$

Reject H_o if $F \geq F_{.05,2,21} = 3.47$

$F = \frac{1856.711/2}{1932.874/21} = 9.98 > 3.47 \Rightarrow$

Reject H_o and conclude there is significant difference in the average hours of relief for the three treatments.

b. 95% C.I. on μ_A : $6.46 \pm (2.080)(9.59)/\sqrt{8} \Rightarrow (-0.592, 13.512)$

95% C.I. on μ_B : $12.34 \pm (2.080)(9.59)/\sqrt{8} \Rightarrow (5.288, 19.392)$

95% C.I. on μ_C : $27.35 \pm (2.080)(9.59)/\sqrt{8} \Rightarrow (20.298, 34.402)$

c. $H_o : \mu_A = \mu_B = \mu_C$ versus H_a : Difference in $\mu's$

Reject H_o if $F \geq F_{.05,2,21} = 3.47$

Using the transformed data: $F = \frac{7.29/2}{7.6727/21} = 10.00 > 3.47 \Rightarrow$

Reject H_o and conclude there is significant difference in the average hours of relief for the three treatments. The following C.I.'s are on the mean of the logarithm of the hours of relief $\mu_{i'}$:

95% C.I. on $\mu_{A'}$: $1.75 \pm (2.080)(0.6055)/\sqrt{8} \Rightarrow (1.305, 2.195)$

95% C.I. on $\mu_{B'}$: $2.43 \pm (2.080)(0.6055)/\sqrt{8} \Rightarrow (1.985, 2.875)$

95% C.I. on $\mu_{C'}$: $3.10 \pm (2.080)(0.6055)/\sqrt{8} \Rightarrow (2.655, 3.545)$

d. The test of hypotheses using the raw and transformed yielded the same conclusion.

e. The inverse transformations involves exponentiating the endpoints of the C.I.:

95% C.I. on μ_A : $(3.688, 8.980)$

95% C.I. on μ_B : $(7.279, 17.725)$

95% C.I. on μ_C : $(14.225, 34.640)$

There is a considerable difference in the two sets of C.I.'s.

8.6 A Nonparametric Alternative: The Kruskal-Wallis Test

8.7 The F-test is testing for a difference in the mean yields of the five varieties; whereas the Kruskal-Wallis test is testing whether there is a difference in the distribution of yields for the five varieties.

8.8   a. The Kruskal-Wallis yields $H = 21.32 > 9.21$ with $df = 2 \Rightarrow p - value < 0.001$. Thus, reject H_o and conclude there is a significant difference in the distributions of deviations for the three suppliers.

b. The box plots and normal probability plot are given here:

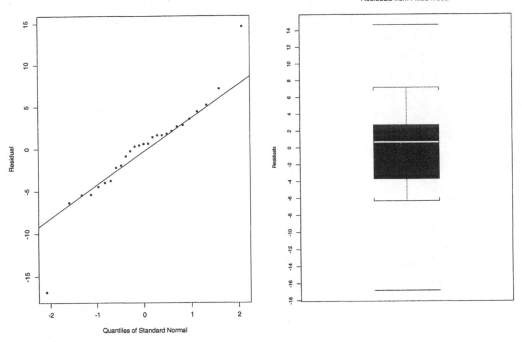

The box plots and normal probability plots of the residuals indicate that the normality condition may be violated. The Levine test yields $L = 3.89$ with $0.025 < p - value < 0.05$. Thus, there is significant evidence that the equal variance condition is also violated

c. The AOV table is given here:

Source	df	SS	MS	F	p-value
Supplier	2	10723.8	5361.9	161.09	0.000
Error	24	798.9	33.3		
Total	26	11522.7			

Reject H_o if $F \geq 3.40$

$F = 161.09 > 3.40$, reject H_o and conclude there is a significant difference in the mean deviations of the three suppliers.

d. 95% C.I. on μ_A : $189.23 \pm (2.064)(5.77)/\sqrt{9} \Rightarrow (185.26, 193.20)$

   95% C.I. on μ_B : $156.28 \pm (2.064)(5.77)/\sqrt{9} \Rightarrow (152.31, 160.25)$

   95% C.I. on μ_C : $203.94 \pm (2.064)(5.77)/\sqrt{9} \Rightarrow (199.97, 207.91)$

Since the upper bound on the mean for supplier B is more than 20 units less than the lower bound on the mean for suppliers A and B, there appears to be a practical difference in the three suppliers. However, because the normality and equal variance assumptions may not be valid, the C.I.'s may not be accurate.

72

Supplementary Exercises

8.9 a. The summary statistics are given here for the observed data:

Line	n	Mean	Std. Dev.
1	5	39.40	7.92
2	5	44.20	10.40
3	5	52.00	26.00
4	5	40.80	10.13
5	5	53.20	17.48

$F_{max} = \frac{(26.00)^2}{(7.92)^2} = 10.78 < 59$ (value from Table 12 with $\alpha = .01$) $\Rightarrow$ There is not significant evidence of a difference in the 5 variances.

b. The summary statistics are given here for the transformed data:

Line	n	Mean	Std. Dev.
1	5	6.25	0.619
2	5	6.61	0.784
3	5	6.93	2.228
4	5	6.35	0.791
5	5	7.22	1.172

$F_{max} = \frac{(2.228)^2}{(0.619)^2} = 12.96 < 59$ (value from Table 12 with $\alpha = .01$) $\Rightarrow$ There is not significant evidence of a difference in the 5 variances for the transformed data.

c. The observed data did not appear to violate the equal variance assumption. Thus, the ANOVA F-test is performed:

$F = \frac{808/4}{5016/20} = 0.81$ with $df = 4, 20 \Rightarrow p - value > 0.25 \Rightarrow$

There is not significance evidence of a difference in the mean number of defects for the five production lines.

8.10 The Kruskal-Wallis test yields $H = 4.32$ with $df = 4 \Rightarrow 0.10 < p - value < 0.90 \Rightarrow$

there is significant evidence of a difference in the distributions of number of defects for the five production lines.

Thus, the Kruskal-Wallis test yields a comparable result as the ANOVA F-test.

8.11 a. $\bar{y}_{..} = \frac{30(90.2)+30(89.3)+30(85)}{90} = \frac{264.5}{3} = 88.1667$

SSB $= 30[(90.2 - \frac{264.5}{3})^2 + (89.3 - \frac{264.5}{3})^2 + (85.0 - \frac{264.5}{3})^2] = 463.40$

SSW $= 29[(6.5)^2 + (7.8)^2 + (7.4)^2] = 4577.65$

F $= \frac{463.40/2}{4577.65/87} = 4.40$ with df $= 2,87 \Rightarrow 0.01 < p - value < 0.025 < 0.05 \Rightarrow$

There is significant evidence of a difference in the mean yields.

b. 95% C.I. on $\mu_1 : 90.2 \pm 1.988 \frac{\sqrt{4577.65/87}}{\sqrt{30}} = (87.57, 92.83)$

95% C.I. on $\mu_2 : 89.3 \pm 1.988 \frac{\sqrt{4577.65/87}}{\sqrt{30}} = (86.67, 91.93)$

95% C.I. on $\mu_3 : 85.0 \pm 1.988 \frac{\sqrt{4577.65/87}}{\sqrt{30}} = (82.37, 87.63)$

c. Although the mean for the control has the lowest value, note that the three C.I.'s have overlap. In Chapter 9, there is a discussion on how to more precisely determine differences in the treatment means.

8.13 a. The Kruskal-Wallis test yields: $H' = 26.62$ with df $= 3 \Rightarrow p - value < 0.001 \Rightarrow$ Thus, there is significant evidence that the distribution of ratings differ for the four groups.

   b. The two procedures yield equivalent conclusions.

8.19 $F = \frac{576.7/2}{1431.6/33} = 6.65$ with df $= 2,33 \Rightarrow 0.001 < p - value < 0.005 < 0.05 \Rightarrow$

There is significant evidence of a difference in the mean lengths of calls for the three areas.

8.21 a. $F = \frac{4020.0/3}{881.9/36} = 54.70$ with df $= 3,36 \Rightarrow p - value < 0.001 < 0.05 \Rightarrow$

There is significant evidence of a difference in the average leaf size under the four growing conditions.

   b. 95% C.I. on μ_A : $23.37 \pm 2.028 \frac{\sqrt{881.9/36}}{\sqrt{10}} = (20.20, 26.54)$

   95% C.I. on μ_B : $8.58 \pm 2.028 \frac{\sqrt{881.9/36}}{\sqrt{10}} = (5.41, 11.75)$

   95% C.I. on μ_C : $14.93 \pm 2.028 \frac{\sqrt{881.9/36}}{\sqrt{10}} = (11.76, 18.10)$

   95% C.I. on μ_D : $35.35 \pm 2.028 \frac{\sqrt{881.9/36}}{\sqrt{10}} = (32.18, 38.52)$

   The C.I. for the mean leaf size for Condition D implies that the mean is much larger for Condition D than for the other three conditions.

   c. $F = \frac{18.08/3}{103.17/36} = 2.10$ with df $= 3,36 \Rightarrow 0.05 < 0.10 < p - value < 0.25 \Rightarrow$

   There is not significant evidence of a difference in the average nicotine content under the four growing conditions.

   d. From the given data, it is not possible to conclude that the four growing conditions produce different average nicotine content.

   e. No. If the testimony was supported by this experiment, then the test conducted in part (c) would have had the opposite conclusion.

8.25 The Kruskal-Wallis test yields $H' = 25.54$ with df $= 4 \Rightarrow p - value < 0.001$

There is significant evidence of a difference in the distribution of the weight gains under the five diets.

The two procedures yield similar conclusions.

8.27 a. The equal variance condition has been violated.

   b. $F = \frac{142.08/4}{42.88/25} = 20.71$ with df $= 4,25 \Rightarrow p - value < 0.001 < 0.05 \Rightarrow$

   The value of the F-statistic has been greatly reduced because of the unequal variance problem.

8.31 The Kruskal-Wallis test yields identical results for the transformed and original data because the transformation was strictly increasing which maintains the order of the data after the transformation has been performed.

   $H = 9.89$ with df $= 2 \Rightarrow 0.005 < p - value < 0.01 < 0.05$ using the Chi-square table.

   Thus, our conclusion is the same as was reached using the transformed data.

Chapter 9: Multiple Comparisons

9.2 Linear Contrasts

9.1 a. Yes, both satisfy +1+1-2=0

b. No, $(+1)(+1) + (+1)(+1) + (-2)(0) + (0)(-2) = 2 \neq 0$

9.3 a. $l_1 = 4\mu_1 - \mu_2 - \mu_3 - \mu_4 - \mu_5$

b. $l_2 = 3\mu_2 - \mu_3 - \mu_4 - \mu_5$

c. $l_3 = \mu_3 - 2\mu_4 + \mu_5$

d. $l_4 = \mu_3 - \mu_5$

9.5 The decision rule for testing $H_o : l = 0$ vs $H_a : l \neq 0$ is

Reject H_o if $F = \frac{SSC}{MS_{Error}} > 4.11$ $(F_{0.05}, df_1 = 1, df_2 = 36)$

$SSC = \frac{(\hat{l})^2}{\sum_i(a_i^2/n_i)} = \frac{10(\hat{l})^2}{\sum_i a_i^2}$

Using the transformed data, $MS_{Error} = 0.2619$

a. $\hat{l}_1 = (1.54) + (2.19) + (4.62) - 3(5.62) = -8.51 \Rightarrow SSC_1 = \frac{10(-8.51)^2}{12} = 60.35 \Rightarrow F = \frac{60.35}{0.2619} = 230.43 \Rightarrow$

Reject H_o and conclude there is significant evidence that the mean oxygen content at 20 KM is different from the average of the mean oxygen content at 1 KM, 5 KM, and 10 KM.

b. $\hat{l}_2 = (1.54) + (2.19) - 2(4.62) = -5.51 \Rightarrow SSC_2 = \frac{10(-5.51)^2}{6} = 50.60 \Rightarrow F = \frac{50.60}{0.2619} = 193.20 \Rightarrow$

Reject H_o and conclude there is significant evidence that the mean oxygen content at 10 KM is different from the average of the mean oxygen content at 1 KM and 5 KM.

c. $\hat{l}_3 = (1.54) - (2.19) = -0.65 \Rightarrow SSC_1 = \frac{10(-0.65)^2}{2} = 2.113 \Rightarrow F = \frac{2.113}{0.2619} = 8.07 \Rightarrow$
Reject H_o and conclude there is significant evidence that the mean oxygen content at 1 KM is different from the mean oxygen content at 5 KM.

d. Yes, because

For $l_1\&l_2 : \sum_i a_i b_i : (1)(1) + (1)(1) + (1)(-2) + (-3)(0) = 0$

For $l_1\&l_3 : \sum_i a_i b_i : (1)(1) + (1)(-1) + (1)(0) + (-3)(0) = 0$

For $l_2\&l_3 : \sum_i a_i b_i : (1)(1) + (1)(-1) + (-2)(0) + (0)(0) = 0$

e. Yes; $SSC_1 + SSC_2 + SSC_3 = 60.35 + 50.60 + 2.11 = 113.06 = SS_{Trt}$

Supplementary Exercises

9.7 The box plots indicate the distribution of the residuals is slightly right skewed. This is confirmed with an examination of the normal probability plot. The Hartley test yields $F_{max} = 2.35 < 7.11$ using an $\alpha = 0.05$ test. Thus, the conditions needed to run the ANOVA F-test appear to be satisfied. From the output, $F = 15.68$, with p-value $< 0.0001 < 0.05$. Thus, we reject H_o and conclude there is significant evidence of a difference in the average weight loss obtained using the five different agents.

9.8   a. Fisher's LSD: Pairs Not Significantly Different: (4,1), (2,3)

   b. Tukey's W: Pairs Not Significantly Different: (4,1), (4,2), (1,2), (2,3), (3,S)

   c. SNK procedure: Pairs Not Significantly Different: (4,1), (2,3)

9.9   a. Tukey's W

   b. Fisher's LSD

9.10 Using the computer output, the Dunnett procedure indicates that all four agents had significantly larger average weight loss than the standard agent.

9.11   a. $l_1 = \mu_{A_1} + \mu_{A_2} + \mu_{A_3} + \mu_{A_4} - 4\mu_S$

   b. $l_2 = \mu_{A_1} - \mu_{A_2} + \mu_{A_3} - \mu_{A_4}$

   c. $l_3 = \mu_{A_1} + \mu_{A_2} - \mu_{A_3} - \mu_{A_4}$

   d. $l_4 = \mu_{A_1} + \mu_{A_3} - 2\mu_S$

9.12 Using Scheffe's Method:

$$S_i = \sqrt{\hat{V}(\hat{l})}\sqrt{(5-1)F_{.05,4,45}} = \sqrt{(.982378)\sum_i a_i^2/10}\sqrt{10.315} = 1.0066\sqrt{\sum_i a_i^2}$$

Declare contrast l_i significantly different from 0 if $|\hat{l}_i| > S_i$

The tests are summarized as follows:

Contrast	a_1	a_2	a_3	a_4	a_5	$\sum_i a_i^2$	$\hat{l}$	S_i	Conclusion
1	1	1	1	1	-4	20	8.5	4.502	Significant
2	1	-1	1	-1	0	4	-.94	2.013	Not Significant
3	1	1	-1	-1	0	4	.56	2.013	Not Significant
4	1	0	1	0	-2	6	3.78	2.466	Significant

We have significant evidence ($\alpha = 0.05$) that the average weight loss for those programs with an agent exceeds the average weight loss for the program using the standard. Also, there is significant evidence that the agents which include counseling yield a higher average weight loss than the program using the standard.

9.13   a. Using Dunnett's 1-sided procedure with LowTar designated the Control Treatment and $\alpha = 0.05$:

$$D = 2.18\sqrt{\tfrac{2(0.159)}{100}} = 0.123 \Rightarrow$$

LowTar has significantly lower average tar content than all four of the other brands.

b. Bonferroni 95% C.I.'s on the differences $\mu_i - \mu_{LowTar}$ with $t_{495,\,\frac{.05}{8}} \approx z_{.0063} = 2.50$, we obtain $(\bar{y}_i - \bar{y}_{LowTar}) \pm 2.50\sqrt{2(.159)/100}$

Brand	C.I.
A	(.436, .718)
B	(.990, 1.272)
C	(1.785, 2.067)
D	(3.807, 4.089)

Note that 0 is not included in any of the four C.I.'s.

9.17　a. Using Fisher's LSD we obtain

Comparison	LSD	$|\bar{y}_i - \bar{y}_j|$	Conclusion
3DOK1 vs 3DOK5	4.501	7.120	Sign. Evid. Means Differ
3DOK1 vs 3DOK7	4.352	8.667	Sign. Evid. Means Differ
3DOK5 vs 3DOK7	4.135	1.557	Not Sign. Evid. Means Differ

b. Based on the normal probability plot of the residuals, it would appear that the residuals may not have a normal distribution since there appears to be several outliers.

Levene's test: $L = 1.57 \Rightarrow p - value = 0.240 \Rightarrow$ There is not significant evidence that the variances are different.

9.24　a. The p-value from the F-test is 0.0345 which is less than 0.05; hence there is significant evidence that the mean fat content in the four treatment groups are different.

b. Using a 1-sided Dunnett's procedure:

Comparison	D	$\bar{y}_i - \bar{y}_A$	Conclusion
B vs A	0.229	0.283	Sign. Evid. B's Mean Differs From Control
C vs A	0.229	0.194	Not Sign. Evid. C's Mean Differs From Control
D vs A	0.229	0.287	Sign. Evid. D's Mean Differs From Control

9.25　Exclude the data from the control group, Treatment A, then combine the data from the three treatment groups into a single data set with a new treatment group designation: Examination Time. Finally, run an ANOVA F-test with the Treatment being Examination Time in order to determine if there is a difference in the mean fat content for the four treatment times. Furthermore, a plot of the treatment means versus Examination Time would would display an increasing trend in the means as the Examination Time increased if there was a time effect. A Bonferroni 1-sided t-test could be run to confirm if the mean fat content was larger for each subsequent Examination Time.

9.27　a. Undergraduate students may not have the proper experience necessary to accurately rate the applicants.

b. Using the same actor will reduce some possible sources of variation. However, the test may be biased if the actor was giving a poor performance, for example.

c. A computer sales position may be easier for some handicapped individuals than others. Varying the type of job being sought would improve the test because the results could provide information concerning a wider variety of situations.

Chapter 10: Categorical Data

10.2 Inferences about a Population Proportion: π

10.1   a. Yes, because $n\hat{\pi} = 30 > 5$ and $n(1 - \hat{\pi}) = 120 > 5$. Samples with $n < 25$ would be suspect.

   b. $.2 \pm 1.645\sqrt{(.2)(.8)/150} \Rightarrow (0.15, 0.25)$ is a 90% C.I. for π.

10.3   a. $\hat{\pi} = 1200/1500 = 0.8 \Rightarrow$ 95% C.I. for $\pi : 0.8 \pm 1.96\sqrt{(.8)(.2)/1500} \Rightarrow (0.780, 0.820)$

10.5   a. Yes, the binomial assumptions hold. The samples are independent, the trials are identical, the probability of success remains constant, and there are two possible outcomes.

   b. Yes, $\pi = \frac{1}{3}, n = 50 \Rightarrow n\pi = 16\frac{2}{3} > 5$ and $n(1 - \pi) = 33\frac{1}{3} > 5$

   c. $\hat{\pi} = \frac{19}{50} = 0.38 \Rightarrow$ 95% C.I. for $\pi : 0.38 \pm 1.96\sqrt{(.38)(.62)/50} \Rightarrow (0.245, 0.515)$

   The C.I. is too wide to be very informative since as an estimate of π it provides values from 25% to over 50% for π.

   In order to decrease the width, the sample size would need to be increased.

10.7   a. By grouping the classes into similar type, it might be possible to summarize the data more concisely. Percentages are helpful but would not add to 100% because one adult might use more than one of the remedies. The numerator of the percentage would refer to users of an OTC remedy and the denominator to the number of patients.

   b. A 95% C.I. using the normal approximation requires that both $n\hat{\pi}$ and $n(1 - \hat{\pi})$ exceed 5. This condition would hold in every OTC category except Sprays/Inhalers, Anesthetic throat lozenges, Room vaporizers and Other products.

10.9 The 95% C.I.'s are given here:

Statement	95% C.I. on Proportion
Others Don't Report	$.56 \pm 1.96\sqrt{(.56)(.44)/500} \Rightarrow (0.516, 0.604)$
Government is Careless	$.50 \pm 1.96\sqrt{(.50)(.50)/500} \Rightarrow (0.456, 0.544)$
Cheating can be Overlooked	$.46 \pm 1.96\sqrt{(.46)(.54)/500} \Rightarrow (0.416, 0.504)$

10.11   a. A table summarizing the results is given here:

Statement	No	Yes
Understand Radiation	70%	30%
Misconceptions About Space-Rockets	40%	60%
Understand How Telephone Works	80%	20%
Understand Computer Software	75%	25%
Understand Gross National Product	72%	28%

   b. Unmentioned details include a complete list of questions asked and the manner in which they were stated. The article also does not report how the survey was conducted. Thus, the results may be biased if the sample was not selected in a random fashion.

For example, if the questionaire were given by mail, the responses would come only those individuals who were able to read and write. This would bias the results because illiterate people probably understand less about technology than literate ones. Other demographic characteristics of the sample might also bias the results.

10.15 Yes, because $n\hat{\pi} = 84 > 5$ and $n(1 - \hat{\pi}) = 41 > 5$

10.17   a. The normal approximation is valid if both $n\pi$ and $n(1 - \pi)$ are greater than 5. With $n = 1500$, unless π is extremely small ($\pi < 0.0033$) or extremely large ($\pi > 0.9967$)the approximation would be valid.

Meal	95% C.I. on 1978 Proportion	95% C.I. Current Proportion
Breakfast	(0.021, 0.039)	(0.039, 0.061)
Lunch	(0.161, 0.199)	(0.180, 0.220)
Dinner	(0.141, 0.179)	(0.141, 0.179)

   b. Since the C.I.'s for 1978 and Now are overlapping, there is not significant evidence of an increase in the proportion of people who are eating meals away from home.

10.19   a. $\hat{\pi} = 560/1500 = 0.373 \Rightarrow$ 95% C.I. on π : $(0.349, 0.397)$
Half width of C.I. is 0.024

   b. Using $\hat{\pi} = 0.373$, $n = \frac{(1.96)^2(0.373)(.627)}{(0.01)^2} = 8984.4 \Rightarrow n = 8985$

10.23   a. $\hat{\pi}_{Adj.} = \frac{\frac{3}{8}}{100 + \frac{3}{4}} = 0.00372$

   b. 99% C.I. on π : $(0, 1 - (.005)^{1/100}) = (0, 0.0516)$

   c. $H_o : \pi \geq 0.01$ versus $H_a : \pi < 0.01$
Because 0.01 falls in the C.I., fail to reject H_o and conclude that the data fails to support the company's claim. The level of the test is $\alpha = \frac{1-.99}{2} = 0.005$. The problem is that with such a small value for π_o, the sample size must be much larger in order for the company to be able to support its claim.

10.3 Inferences about the Difference between $\pi_1 - \pi_2$

10.27 Because p-value $= 0.0218 < 0.05$, reject H_o and conclude that the data supports the hypothesis that the rates of satisfied customers served by the two methods are different.

10.28 95% C.I. on $\pi_1 - \pi_2$: $(0.013, 0.167)$

Because 0 is not contained within the C.I., H_o is rejected and hence the conclusion is the same as was in Exercise 10.27.

10.31   a. $\hat{\pi}_1 = 22/30 = .73, \hat{\pi}_2 = 16/30 = .53 \Rightarrow \hat{\sigma} = \sqrt{\frac{.73(1-.73)}{30} + \frac{.53(1-.53)}{30}} = 0.121293 \Rightarrow$
$z = \frac{.73-.53}{0.121293} = 1.64 \Rightarrow$ p-value $= 2Pr(z > 1.64) = 0.101 > 0.05 \Rightarrow$ Fail to reject H_o.
There is not significant evidence that the reliabilities are different.

b. No, a failure to reject H_o has only demonstrated there is not significant evidence to state there is a difference in the reliabilities. We would need to determine the probability of Type II error for various values of $\pi_1 - \pi_2$ in order to be confident in making a statement concerning whether the reliabilities may be the same.

10.32 $\hat{\pi}_1 - \hat{\pi}_2 = \frac{6}{30} = 0.2 \Rightarrow$ 95% C.I. on $\pi_1 - \pi_2$:

$0.2 \pm (1.96)(0.121293) \Rightarrow (-0.038, 0.438) \Rightarrow$

We are 95% confident that the difference in the reliabilities of the two suppliers is between -0.038 and 0.438. Because the interval contains 0, we are unable to state that the reliabilities of the two suppliers differ.

10.33 $\hat{\pi}_1 = \frac{25}{100} = .25,\quad \hat{\pi}_2 = \frac{15}{100} = .15 \Rightarrow \hat{\pi}_1 - \hat{\pi}_2 = 0.1 \Rightarrow$ 95% C.I. on $\pi_1 - \pi_2$:

$0.1 \pm (1.96)\sqrt{\frac{0.25(1-0.25)}{100} + \frac{0.15(1-0.15)}{100}} \Rightarrow (-0.01, 0.21)$

10.35 a. $z = \dfrac{0.90-0.36}{\sqrt{\frac{0.9(1-0.9)}{100} + \frac{0.36(1-0.36)}{100}}} = 9.54 \Rightarrow p-value = Pr(z > 9.54) < 0.0001$

Reject H_o and conclude there is significant evidence that the death rate after 30 days is greater for Cocaine group than for the Heroin group.

b. If the physical response to the two drugs is the same for humans, cocaine is a very dangerous drug, even more so than heroin.

10.4 Chi-Square Goodness-of-Fit Test

10.43 $H_o : \pi_1 = \frac{1}{3}, \pi_2 = \frac{1}{3}, \pi_3 = \frac{1}{3}$

H_a : at least on of the groups had probability of interning different from $\frac{1}{3}$

$E_i = n\pi_{io} \Rightarrow\quad E_1 = \frac{63}{3} = 21,\quad E_2 = \frac{63}{3} = 21\quad E_3 = 63\frac{63}{3} = 21$

$\chi^2 = \sum_{i=1}^{3} \frac{(n_i - E_i)^2}{E_i} = 6.952$ with $df = 3 - 1 = 2 \Rightarrow 0.025 < p-value < 0.05 \Rightarrow$ p-value $> 0.01 \Rightarrow$ Fail to reject H_o.

There is not substantial evidence that the probability that a student interned in one of the three industries will finish the program differs from 1/3.

10.45 $H_o : \pi_1 = 0.50, \pi_2 = 0.40, \pi_3 = 0.10$

H_a : at least on of the $\pi_i s$ differs from its hypothesized value

$E_i = n\pi_{io} \Rightarrow\quad E_1 = 200(.5) = 100,\quad E_2 = 200(.4) = 80,\quad E_3 = 200(.1) = 20$

$\chi^2 = \sum_{i=1}^{3} \frac{(n_i - E_i)^2}{E_i} = 6.0$ with $df = 3 - 1 = 2 \Rightarrow 0.025 < p-value < 0.05 \Rightarrow$

Reject H_o at the $\alpha = 0.05$ level. There is substantial evidence that the distribution of registered voters is different from previous elections.

10.47 If there was no seasonal trend, we would expect colds to occur with equal probability in each of the four seasons, that is, $\pi_i = \frac{1}{4} = .25$.

$H_o : \pi_1 = 0.25, \pi_2 = 0.25, \pi_3 = 0.25, \pi_4 = 0.25$

H_a : at least on of the $\pi_i s$ differs from 0.25

$E_i = n\pi_{io} \Rightarrow \quad E_1 = 1000(.25) = 250, \quad E_2 = 1000(.25) = 250,$

$E_3 = 1000(.25) = 250, \quad E_4 = 1000(.25) = 250$

$\chi^2 = \sum_{i=1}^{4} \frac{(n_i - E_i)^2}{E_i} = 123.70$ with $df = 4 - 1 = 3 \Rightarrow p - value < 0.001 \Rightarrow$

Reject H_o. There is substantial evidence that there is a seasonal trend in the occurrence of colds.

10.49 $H_o : \pi_1 = 0.25, \pi_2 = 0.25, \pi_3 = 0.25, \pi_4 = 0.25$

H_a : at least on of the $\pi_i s$ differs from its hypothesized value

$E_i = n\pi_{io} \Rightarrow \quad E_1 = 40(.25) = 10, \quad E_2 = 40(.25) = 10,$

$E_3 = 40(.25) = 10, \quad E_4 = 40(.25) = 10$

$\chi^2 = \sum_{i=1}^{4} \frac{(n_i - E_i)^2}{E_i} = 2.4$ with $df = 4 - 1 = 3 \Rightarrow p - value > 0.10 \Rightarrow$

Fail to reject H_o. The data supports that claim that the editors' are equally opinioned over the four categories. However, the chance of making a Type II error is unknown and the sample size is very small. Thus, the conclusion should not be given a great deal of crediability.

10.5 The Poisson Distribution

10.51   a. The probabilities are given here: $Pr(y > 1) = 1 - Pr(y \le 1)$

μ	0.5	1.0	3.0
$P(y = 1)$	0.3033	0.3679	0.1494

b. The probabilities are given here: $Pr(y < 5) = Pr(y \le 4)$

μ	1.7	2.5	4.2
$P(y > 1)$	0.5067	0.7127	0.9220

c. The probabilities are given here:

μ	0.2	1.0	2.0
$P(y < 5)$	1.0	0.9964	0.9476

10.53 Let y be the number of sales in the 250 calls. Under the binomial conditions, y has a binomial distribution with parameters, $\pi = 0.01$ and $n = 250$.

a. $P(y < 6) = \sum_{k=0}^{5} \frac{250!}{k!(250-k)!}(0.01)^k(1-0.01)^{250-k}$

b. The 250 calls are independent and identical resulting in either a sale or no sale. The probability of making a sale is constant for all 250 calls.

c. $\pi = 0.01, \quad n = 250, \quad \mu = n\pi = 250(0.01) = 2.5, \quad \sigma = \sqrt{250(0.01)(1-0.01)} = 1.5732$

$P(y < 6) \approx P(z < \frac{5.5-2.5}{1.5732}) = P(z < 1.91) = 0.9719$

d. $\mu = n\pi = 2.5 \Rightarrow$

$P(y < 6) \approx \sum_{k=0}^{5} P(y = k) = 0.0821 + 0.2052 + 0.2565 + 0.2138 + 0.1336 + 0.0668 = 0.9580$

e. Because $n\pi = 2.5 < 5$, the Poisson gives a better approximation. (In fact, using a computer program for the binomial distribution, $P(y < 6) = 0.9588$.)

10.55 a. Yes, the Poisson conditions appear to be reasonably satisfied in this situation.

b. $H_o : \mu = 2$ versus $H_a : \mu \neq 2$

Using $\mu = 2.0$, the Poisson table yields the following probabilities:

k	0	1	2	3	4	5	≥ 6
$\pi_i = P(y = k)$	0.1353	0.2707	0.2702	0.1804	0.0902	0.0361	0.0166
$E_i = 800\pi_i$	108.24	216.56	216.56	144.32	72.16	28.88	13.28

$\chi^2 = \sum_{i=1}^{7} \frac{(n_i - E_i)^2}{E_i} = 8.272$ with $df = 7 - 1 = 6 \Rightarrow p - value > 0.10 \Rightarrow$
Fail to reject H_o. There is not significant evidence to reject the claim that number of conflicts per 5 minutes differs from 2.

10.57 a. From the data $\bar{y} \approx \frac{1}{100} \sum_i (n_i)(y_i) = 5.57$

$s^2 \approx \frac{1}{99} \sum_i n_i(y_i - 5.57)^2 = 1056.5/99 = 10.67$

b. Using $\mu = 5.5$, the Poisson table yields the following probabilities after combining the first 2 categories and combining the last four categories so that $E_i > 1$ and only one E_i is less than 5:

k	≤ 1	2	3	4	5	6
$\pi_i = P(y = k)$	0.0266	0.0618	0.1133	0.1558	0.1714	0.1571
$E_i = 100\pi_i$	2.66	6.18	11.33	15.58	17.14	15.71
k	7	8	≥ 9			
$\pi_i = P(y = k)$	0.1234	0.0849	0.1057			
$E_i = 100\pi_i$	12.34	8.49	10.57			

$\chi^2 = \sum_{i=1}^{9} \frac{(n_i - E_i)^2}{E_i} = 13.441$ with $df = 9 - 2 = 7 \Rightarrow 0.05 < p - value < 0.10 \Rightarrow$
Fail to reject H_o. The conclusion that the number of fire ant hills follows a Poisson distribution appears to be supported by the data. However, we have not computed the probability of making a Type II error so the conclusion is somewhat tenuous.

c. The fire ant hills are somewhat more clustered than randomly distributed across the pastures, although the data failed to reject the null hypothesis that the fire ant hills were randomly distributed.

10.6 Contingency Tables

10.60 a. The expected counts are $E_{ij} = n_{i.}n_{.j}/250$ and are displayed in the following table:

		Column		
Row	1	2	3	4
1	16.0	22.4	25.6	16.0
2	34.0	47.6	54.4	34.0

b. df = (2-1)(4-1) = 3

c. Using the chi-square approximation with df = 3 and $\chi^2 = 13.025$,
  $\Rightarrow 0.001 < p - value < 0.005$. Thus, there is significant evidence of a relationship.

10.61 $0.001 < p - value < 0.005$

Supplementary Exercises

10.67 a. Under the hypothesis of independence, the expected frequencies are given in the following table with $\hat{E}_{ij} = n_{i.}n_{.j}/900$

		Opinion			
Commercial	1	2	3	4	5
A	43	107	78	34	39
B	42	107	78	34	39
C	42	107	78	34	39

b. df = (3-1)(5-1) = 8

c. The cell chi-squares are given in the following table:

		Opinion			
Commercial	1	2	3	4	5
A	2.3810	3.7383	2.1667	4.2353	0.6410
B	2.8810	10.8037	0.0513	5.7647	21.5641
C	0.0238	1.8318	1.5513	0.1176	14.7692

$\chi^2 = \sum_{i,j} \frac{(n_{ij}-E_{ij})^2}{E_{ij}} = 72.521$ with $df = 8 \Rightarrow p - value = P(\chi^2 > 72.521) < 0.001 \Rightarrow$
Reject H_o. There is significant evidence that the Commercial viewed and Opinion are related.

10.68 With $df = 8, p - value = P(\chi^2 > 72.521) < 0.001$

10.69 Since interst lies in understanding the relationship between Opinion rating given the Commercial viewed, we will designate Commercial as the independent variable and Opinion as the dependent variable. Therefore, we will try to predict Opinion as a function of Commercial. The relevant proportions are given in the following table:

| Commercial | Opinion | | | | | Total |
	1	2	3	4	5	
A	0.1067	0.2900	0.3033	0.1533	0.1467	1.0
B	0.1767	0.4700	0.2533	0.0667	0.0333	1.0
C	0.1367	0.3100	0.2233	0.1200	0.2100	1.0
Total	0.1400	0.3567	0.2600	0.1133	0.1300	1.0

When Commercial A is viewed, the most common Opinion is 3, meaning $32+87+46+44 = 209$ prediction errors are made. When Commercial B is viewed, the most common Opionion is 2, leaving $53+76+20+10=159$ prediction errors. When Commercial C is viewed, the most common Opinion is 2, yielding $41+67+36+63=207$ prediction errors. The total number of errors is thus $209+159+207=575$.

If the Commercial is unknown, we would predict an Opinion of 2 (with 321 votes). This leaves $126+234+102+117=579$ errors.

Therefore, $\lambda=(579-575)/579=0.0069$. We would make 0.69% fewer errors predicting Opinion with knowledge of the Commercial viewed than without this knowledge. This is not a strong relation.

10.73   a. For selection rating 1, there are 11 responses ($11/216=5.09\%$). For selection rating 2, there are 30 responses ($30/216=13.89\%$). For selection rating 3, there are 88 responses ($88/216=40.74\%$). Finally, for selection rating 4, there are 87 responses ($87/216=40.28\%$).

   b. The goodness-of-fit test is computed in the following table:

Selection Rating	Theoretical Proportions π_i	Expected Frequencies $E_i = 216\pi_i$	Observed Frequencies n_i	Cell chi-square: $(n_i - E_i)^2/E_i$
1	0.25	54	11	34.2407
2	0.25	54	30	10.6667
3	0.25	54	88	21.40474
4	0.25	54	87	20.1667
Total	1.00	216	216	86.4815

$H_o : \pi_1 = 0.25, \pi_2 = 0.25, \pi_3 = 0.25, \pi_4 = 0.25$ versus

H_a : Selection Rating categories are not equally likely

$\chi^2 = \sum_{i=1}^4 \frac{(n_i - E_i)^2}{E_i} = 86.4815$ with $df = 4 - 1 = 3 \Rightarrow p - value < 0.001 \Rightarrow$
Reject H_o. There is significant evidence that the selection rating categories are not equally likely.

10.74   a. The initial analysis of the data yielded a contingency table with 5 of the 16 cells (31%) having expected values less than 5. The first two levels of the adequacy rating were combined. The following SAS output was then obtained.

SAS System:

The FREQ Procedure

Table of Adequacy Rating by Frequency

```
Adequacy Rating|      Frequency
               |
Frequency      |
Expected       |
Cell Chi-Square|
Percent        |
Row Pct        |       |       |       |       |
Col Pct        |      1|      2|      3|      4|  Total
---------------+-------+-------+-------+-------+
            1 |     5 |     8 |    11 |    17 |     41
              |16.324 |12.718 |7.5926 |4.3657 |
              |7.8556 |  1.75 |1.5292 |36.563 |
              |  2.31 |  3.70 |  5.09 |  7.87 |  18.98
              | 12.20 | 19.51 | 26.83 | 41.46 |
              |  5.81 | 11.94 | 27.50 | 73.91 |
---------------+-------+-------+-------+-------+
            3 |    37 |    30 |    16 |     5 |     88
              |35.037 |27.296 |16.296 |9.3704 |
              |  0.11 |0.2678 |0.0054 |2.0384 |
              | 17.13 | 13.89 |  7.41 |  2.31 |  40.74
              | 42.05 | 34.09 | 18.18 |  5.68 |
              | 43.02 | 44.78 | 40.00 | 21.74 |
---------------+-------+-------+-------+-------+
            4 |    44 |    29 |    13 |     1 |     87
              |34.639 |26.986 |16.111 |9.2639 |
              |2.5298 |0.1503 |0.6008 |7.3718 |
              | 20.37 | 13.43 |  6.02 |  0.46 |  40.28
              | 50.57 | 33.33 | 14.94 |  1.15 |
              | 51.16 | 43.28 | 32.50 |  4.35 |
---------------+-------+-------+-------+-------+
Total         |    86      67      40     23 |    216
              | 39.81   31.02   18.52   10.65 | 100.00
```

Statistics for Table of Adequacy Rating by Frequency

Statistic	DF	Value	Prob
Chi-Square	6	60.7719	<.0001
Likelihood Ratio Chi-Square	6	53.1543	<.0001
Mantel-Haenszel Chi-Square	1	46.7048	<.0001
Phi Coefficient		0.5304	
Contingency Coefficient		0.4686	
Cramer's V		0.3751	

Sample Size = 216

The contingency table now has only 1 cell with expected count less than 5. The Pearson Chi-square value is 60.7719 with df = 6 and p-value < 0.0001. Thus, the null hypothesis that frequency of use and adequacy rating are independent is rejected. We thus conclude that there is some relation between the two variables.

b. Before combining the two lowest levels of adequacy rating 31% of the expected counts were less than 5. However, after combining the cells, only 1 of the 12 cells (8.3%) had expected count less than 5. Thus, the chi-square approximation should yield a reasonably accurate p-value for the test of independence.

10.75 The counts and adequacy rating percentages (column percentages) for each frequency level are given in the table contained in the solution for Exercise 10.74. It appears that when frequency use is lower, the adequacy ratings tend to be higher. For instance, when frequency use is 1, over 94% of respondents gave a rating or 3 or 4. Although there is slight drop off, low frequency users tend to give a high selection rating. This trend even carries over into frequency use category 3, where 72.5% of respondents gave a 3 or 4 rating. Only in the highest frequency of use category does this trend reverse itself. Only 26.09% of respondents in this group gave a selection rating of 3 or 4, while 73.91% gave a selection rating or 1 or 2. One possible explanation for this result is perhaps these people have rented so many videos that no store could possibly have a good enough selection for them because they have rented just about all the videos in the store.

10.78 a. Control: 10%; Low Dose 14%; High Dose 19%

b. $H_o : \pi_1 = \pi_2 = \pi_3$ versus H_a : The proportions are not all equal,
where π_j is probability of a rat in Group j having One or More Tumors.

$E_{ij} = 100n_{.j}/300$ and $\chi^2 = \sum_{ij} \frac{(n_{ij}-E_{ij})^2}{E_{ij}} = 3.312$ with df = (2-1)(3-1) = 2 and p-value = 0.191.

Because the p-value is fairly large, we fail to reject H_o and conclude there is not significant evidence of a difference in the probability of having One or More Tumors for the three rat groups.

c. No, since we the chi-square test failed to reject H_o.

10.79 Same results.

10.81 a. The results are summarized in the following table: with
$\hat{\sigma}_{\hat{\pi}} = \sqrt{(\hat{\pi})(1-\hat{\pi})/500}$ and 95% C.I. $\hat{\pi} \pm 1.96\hat{\sigma}_{\hat{\pi}}$

Question	$\hat{\pi}$	$\hat{\sigma}_{\hat{\pi}}$	95% C.I.
Did Not Explain?	0.254	0.01947	(0.216, 0.292)
Might Bother?	0.916	0.0124	(0.892, 0.940)
Did Not Ask?	0.471	0.02232	(0.427, 0.515)
Drug Not Changed?	0.877	0.0147	(0.848, 0.906)

b. It would be important to know how the patients were selected, how the questions were phrased, the condition of the illness, and many other factors.

10.85 $\bar{y} = \sum_i (No.Mites)(frequency)/500 = 1.146$

After combining the last 3 categories so that all $E_i > 1$ and only 1 $E_i < 5$, we obtain the following using a Poisson distribution with $\mu = 1.2$:

Mites/Leaf (k_i)	0	1	2	3	4	≥ 5	Total
$\pi_i = P(y = k_i)$	0.3012	0.3614	0.2169	0.0867	0.0260	0.0078	1.0
$E_i = 500\pi_i$	150.6	180.7	108.5	43.3	13	3.9	500
n_i	233	127	57	33	30	20	500

$\chi^2 = \sum_i \frac{(n_i - E_i)^2}{E_i} = 176.6$ with df $=6\text{-}1 = 5, \Rightarrow p - value = Pr(\chi^2 > 176.6) < 0.001 \Rightarrow$ Reject H_o, and conclude there is significant evidence that the data does not fit a Poisson distribution with $\mu = 1.2$.

Chapter 11: Linear Regression and Correlation

11.2 Estimating Model Parameters

11.3 The calculations are give here:

i	x_i	y_i	$(x_i - 3)^2$	$(x_i - 3)(y_i - 5.6)$
1	1	2	4	7.2
2	2	4	1	1.6
3	3	6	0	0
4	4	7	1	1.4
5	5	9	4	6.8
Total	15	28	10	17

$\bar{x} = 15/5 = 3 \qquad \bar{y} = 28/5 = 5.6$

$S_{xx} = \sum_i (x_i - 3)^2 = 10$

$S_{xy} = \sum_i (x_i - 3)(y_i - 5.6) = 17$

$\hat{\beta}_1 = S_{xy}/S_{xx} = 17/10 = 1.7$

$\hat{\beta}_0 = \bar{y} - \hat{\beta}_1 \bar{x} = 5.6 - (1.7)(3) = 0.5$

$\hat{y} = 0.5 + 1.7x$

11.5 The calculations are give here:

i	x_i	y_i	$(x_i - 14)^2$	$(x_i - 14)(y_i - 25.33)$
1	5	10	81	137.97
2	10	19	16	25.32
3	12	21	4	8.66
4	15	28	1	2.07
5	18	34	16	34.68
6	24	40	100	146.70
Total	84	152	218	356

$\bar{x} = 84/6 = 14 \qquad \bar{y} = 152/6 = 25.33$

$S_{xx} = \sum_i (x_i - 14)^2 = 218$

$S_{xy} = \sum_i (x_i - 14)(y_i - 25.33) = 356$

$\hat{\beta}_1 = S_{xy}/S_{xx} = 356/218 = 1.633$

$\hat{\beta}_0 = \bar{y} - \hat{\beta}_1 \bar{x} = 25.33 - (1.633)(14) = 2.468$

$\hat{y} = 2.468 + 1.633x$

11.9   a. A scatterplot of the data is given here:

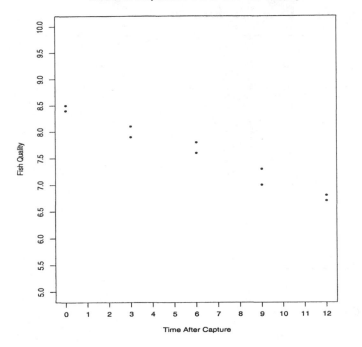

Effect of Exposure Time on Fish Quality

b. Minitab output is given here:

```
Regression Analysis: QUALITY versus STORAGE


The regression equation is
QUALITY = 8.46 - 0.142 STORAGE

Predictor        Coef     SE Coef         T        P
Constant      8.46000     0.06610    127.99    0.000
STORAGE      -0.141667    0.008995    -15.75    0.000

S = 0.1207     R-Sq = 96.9%     R-Sq(adj) = 96.5%

Analysis of Variance

Source           DF        SS        MS        F        P
Regression        1    3.6125    3.6125   248.07    0.000
Residual Error    8    0.1165    0.0146
Total             9    3.7290
```

The least squares estimates are $\hat{\beta}_0 = 8.46$ and $\hat{\beta}_1 = -0.142$

c. The estimated slope value $\hat{\beta}_1 = -0.142$ indicates that for each 1 hour increase in the time between the capture of the fish and their placement in storage there is approximately a 0.142 decrease in the average quality of the fish.

11.10 $\hat{y} = 8.46 - 0.142x$ $\Rightarrow$ $\hat{y} = 8.46 - 0.142(10) = 7.04$

90

It would not be a good idea to make predictions at $x = 18$ because the experimental data contained values between 0 and 12 hours. Thus, 18 hours is well beyond the experimental data.

11.15 The Time Needed is increasing as the Number of Items increases but the rate of increase is less for larger values of Number of Items than it is for smaller values of Number of Items. A square root or logarithmic transformation is suggested from the plot.

11.16   a. For the transformed data, the plotted points appear to be reasonably linear.

     b. The least squares line is $\hat{y} = 3.10 + 2.76\sqrt{x}$

       (Using rounded values $3.097869 \approx 3.10$ and $2.7633138 \approx 2.76$).

       That is, Estimated Time Needed = $3.10 + 2.76\sqrt{\text{Number of Items}}$.

11.17 The Root Mean Square Error (RMSE) is 2.923232. This is the square root of the "average" sum of squares of the y-values about the least squares line. This is an assessment of how "close" the data values are to the least squares line in the y-direction.

11.18 The RMSE for the transformed data can be compared to the RMSE for the original data because only the x-values where transformed. Because the y-values have not been transformed and RMSE measures distance from the regression line and the plotted points in the y-direction, the units for RMSE are the same for both regression lines.

11.21   a. A scatterplot of the data is given here:

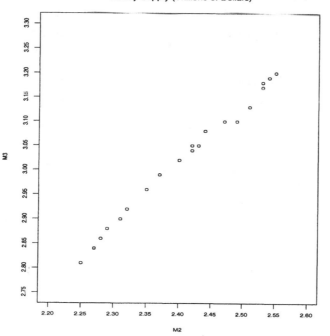

91

From the plot of the data, there is a definite pattern in the lifetime of the bits as the speed is increased. The lifetimes initially increase, then decrease, as the speed is increased. The relationship between lifetime and speed is not a straight-line.

b. The lifetime of one of the bits at a speed 100 is quite a bit smaller than the other three lifetimes at this speed. However, because a speed of 100 is the average of all speeds in the speed, this lifetime has very low leverage and hence can not have high influence.

11.22　a. The estimated intercept is 6.03 and the estimated slope is -0.017.

b. The slope has a negative sign which indicates a decreasing relation between lifetime and speed. The scatterplot indicates that this is misleading. The relation is not a straight-line. In fact, as speed increases, the lifetimes initially increase, then decrease.

c. The residual standard deviation is given as "Standard Error", 0.6324. This value is the square root of the MS(Residual), that is, $0.6324 = \sqrt{0.400}$. This value is the square root of the average squared deviation of the data values about the fitted line.

11.23　a. The prediction equation is $\hat{y} = 6.03 - 0.017x$. Thus, we just replace x with the speeds of 60, 80, 100, 120, and 140 to obtain the values:

$$x = 60 \Rightarrow \hat{y} = 6.03 - (0.017)(60) = 5.01$$

$$x = 80 \Rightarrow \hat{y} = 6.03 - (0.017)(80) = 4.67$$

$$x = 100 \Rightarrow \hat{y} = 6.03 - (0.017)(100) = 4.33$$

$$x = 120 \Rightarrow \hat{y} = 6.03 - (0.017)(120) = 3.99$$

$$x = 140 \Rightarrow \hat{y} = 6.03 - (0.017)(140) = 3.65$$

b. The predicted values are larger than all the observed lifetimes at both the lowest speed, 60 and the highest speed, 140. Most of the lifetime values are greater than the predicted values for speeds 80 to 120. Thus we can conclude that the straight-line model for this data is not appropriate. If the fitted line is reasonable, there should not be any systematic pattern in the deviations of the observed data values from the predictions. An alternative model should be fit to these data values.

11.3 Inferences about Regression Parameters

11.29 Minitab output is given here:

```
Regression Analysis: FIRMNESS versus CONCENTRATION


The regression equation is
FIRMNESS = 48.9 + 10.3 CONCENTRATION

Predictor        Coef      SE Coef         T        P
Constant       48.933        1.541     31.76    0.000
CONCENTR      10.3333        0.7957    12.99    0.000

S = 2.387      R-Sq = 97.7%      R-Sq(adj) = 97.1%

Analysis of Variance

Source            DF          SS         MS        F        P
Regression         1      961.00     961.00   168.65    0.000
Residual Error     4       22.79       5.70
Total              5      983.79
```

    a. $y = 48.9 + 10.3x$

    b. $S^2 = 5.70$

    c. $SE(\hat{\beta}_1) = 0.7957$

11.30 $H_o : \beta_1 = 0$ versus $H_a : \beta_1 \neq 0$

Test Statistic: $t = \frac{\hat{\beta}_1}{s_e/S_{xx}} = \frac{10.33}{2.387/9} = 12.99$

$p-value = 2P(t_4 > 12.99) < 0.0001 \Rightarrow$ Reject H_o and conclude there is significant evidence that β_1 is not 0.

11.35   a. Yes, the data values fall approximately along a straight-line.

    b. $\hat{y} = 12.51 + 35.83x$

11.36   a. 1.069

    b. 6.057

    c. $H_o : \beta_1 \leq 0$ versus $H_a : \beta_1 > 0$

    Test Statistic: $t = 5.15$

    $p - value = P(t_{10} > 5.15) = 0.0002 \Rightarrow$ Reject H_o and conclude there is significant evidence that there is a positive linear relationship.

11.37   a. No because the interpretation for β_o is the average weight gain for chickens who did not ingest lysine in their diet. However, lysine was placed in all the feed. Thus, all chickens consumed some lysine.

    b. The model $y = \beta_1 x + \epsilon$ forces the estimated regression line to pass through the origin, the point (0,0). The model $y = \beta_o + \beta_1 x + \epsilon$ does not necessarily force the fitted line to pass through the origin. This provides greater flexibility in fitting models where there was no data collected near x=0.

11.38   a. $\hat{\beta}_1 = 106.52$

b. Scatterplot of the data and the two fitted lines are given here:

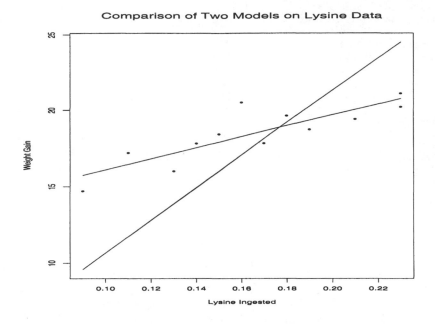

The model with an intercept term produced a better fit to the data values.

11.39   a. Scatterplot of the data is given here:

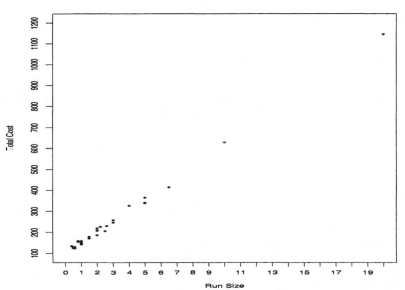

An examination of the scatterplot reveals that a straight-line equation between total cost and run size may be appropriate. There is a single extreme point in the data set but no evidence of a violation of the constant variance requirement.

b. $\hat{y} = 99.777 + 5.1918x$

The residual standard deviation is $s = \sqrt{148.999} = 12.2065$

c. A 95% C.I. for the slope is given by $\hat{\beta}_1 \pm t_{0.025,28} SE(\hat{\beta}_1) \Rightarrow$
$5.1918 \pm (2.048)(0.0586455) \Rightarrow \quad (5.072, 5.312)$

11.4 Predicting New y Values Using Regression

11.43 95% prediction interval for log biological recovery percentage at x=30 is given by

$$\hat{y} \pm t_{.025,11} s_\epsilon \sqrt{1 + \frac{1}{n} + \frac{(30-\bar{x})^2}{S_{xx}}} \Rightarrow$$

$1.195 \pm (2.201)(0.1114)\sqrt{1 + \frac{1}{13} + \frac{(30-30)^2}{4550}} \Rightarrow \quad 1.195 \pm 0.0.254 \Rightarrow \quad (0.941, 1.449)$

The prediction interval is somewhat wider than the confidence interval on the mean.

11.44   a. $\hat{y} = -1.733333 + 1.316667x$

b. The p-value for testing $H_o : \beta_1 \leq 0$ versus $H_a : \beta_1 > 0$ is
$p - value = P(t_{10} \geq 6.342) < 0.0005 \Rightarrow$ Reject H_o and conclude there is significant evidence that the slope β_1 is greater than 0.

11.45   a. 95% Confidence Intervals for $E(y)$ at selected values for x:
$x = 4 \Rightarrow (2.6679, 4.3987)$
$x = 5 \Rightarrow (4.2835, 5.4165)$
$x = 6 \Rightarrow (5.6001, 6.7332)$
$x = 7 \Rightarrow (6.6179, 8.3487)$

b. 95% Prediction Intervals for y at selected values for x :
$x = 4 \Rightarrow (1.5437, 5.5229)$
$x = 5 \Rightarrow (2.9710, 6.7290)$
$x = 6 \Rightarrow (4.2877, 8.0456)$
$x = 7 \Rightarrow (5.4937, 9.4729)$

c. The confidence intervals in part (a) are interpreted as "We are 95% confident that the average weight loss over many samples of the compound when exposed for 4 hours will be between 2.67 and 4.40 pounds." Similar statements for other hours of exposure.

The prediction intervals in part (b) are interpreted as "We are 95% confident that the weight loss of a single sample of the compound when exposed for 4 hours will be between 1.54 and 5.52 pounds." Similar statements for other hours of exposure.

11.46   a. $\hat{y} = 99.77704 + 51.9179x \Rightarrow$ When x=2.0, $E(y) = 99.77704 + (51.9179)(2.0) = 203.613$ as is shown in the output.

b. The 95% C.I. is given in the output as $(198.902, 208.323)$

11.47 No, because $x = 2.0$ is close to the mean of all x-values used in determining the least squares line.

11.48   a. The 95% P.I. is given in the output as $(178.169, 229.057)$

b. Yes, because \$250 does not fall within the 95% prediction interval. In fact, it is considerably higher that the upper value of \$229.057.

11.5 Examining Lack of Fit in Linear Regression

11.53   a. Scatterplot of the data is given here:

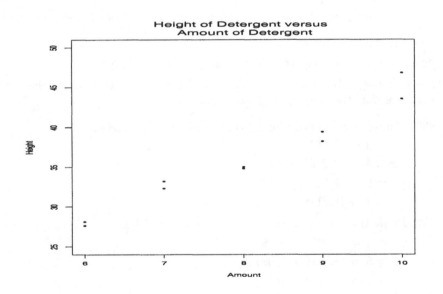

b. The SAS output is given here:

```
Model: MODEL1
Dependent Variable: Y

                         Analysis of Variance

                        Sum of          Mean
    Source      DF      Squares        Square     F Value    Prob>F

    Model        1     330.48450     330.48450    169.213    0.0001
    Error        8      15.62450       1.95306
    C Total      9     346.10900

        Root MSE      1.39752     R-square     0.9549
        Dep Mean     35.89000     Adj R-sq     0.9492
        C.V.          3.89390

                        Parameter Estimates

                    Parameter      Standard    T for H0:
    Variable   DF    Estimate        Error    Parameter=0   Prob > |T|

    INTERCEP    1    3.370000     2.53872138      1.327        0.2210
    X           1    4.065000     0.31249500     13.008        0.0001
```

$\hat{y} = 3.37 + 4.065x$

c. The residual plot is given here:

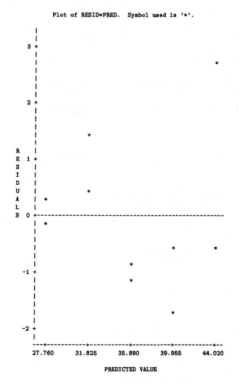

Plot of RESID*PRED. Symbol used is '*'.

The residual plot indicates that higher order terms in x may be needed in the model.

11.54 a. A test of lack of fit is given here:

$SSP_{exp} = \sum_{ij}(y_{ij} - \bar{y}_{i.})^2 = (28.1 - 27.85)^2 + (27.6 - 27.85)^2 + (32.3 - 32.75)^2 + (33.2 - 32.75)^2 + (34.8 - 34.90)^2 + (35.0 - 34.90)^2 + (38.2 - 38.80)^2 + (39.4 - 38.80)^2 + (43.5 - 45.15)^2 + (46.8 - 45.15)^2 = 6.715$

From the output of exercise 11.53, SS(Residuals) = 15.6245. Thus,

$SS_{Lack} = SS(Residuals) - SS_{exp} = 15.6245 - 6.715$

$df_{Lack} = n - 2 - \sum_i(n_i - 1) = 10 - 2 - 5(2-1) = 3$

$df_{exp} = \sum_i(n_i - 1) = 5(2-1) = 5$

$F = \frac{8.9095/3}{6.715/5} = 2.21 < 5.41 = F_{.05,3,5}$

There is not sufficient evidence that the linear model is inadequate.

b. $S_{xx} = \sum_i(x_i - \bar{x}) = 20$. Prediction interval for y at x is

$\hat{y} \pm t_{.025,8} s_\epsilon \sqrt{1 + \frac{1}{n} + \frac{(x-\bar{x})^2}{S_{xx}}} = 3.37 + 4.065x \pm (2.306)(1.398)\sqrt{1 + \frac{1}{10} + \frac{(x-8)^2}{20}}$

For x=6,7,8,9,10 we have

x	$\hat{y}$	95% P.I.
6	27.760	(24.086, 31.434)
7	31.825	(28.369, 35.281)
8	35.890	(32.510, 39.270)
9	39.955	(36.499, 43.411)
10	44.020	(40.346, 47.694)

11.6 The Inverse Regression Problem (Calibration)

11.57 a. $\hat{y} = 0.11277 + 0.11847x \approx 0.113 + 0.118x$

Dependent variable is Transformed CUMVOL and Independent variable is Log(Dose)

b. $\hat{x} = (y - 0.11277)/0.11847$

$s_\epsilon = 0.04597, S_{xx} = 4.9321, \bar{x} = 2.40, t_{.005,8} = 2.819 \Rightarrow$

$c^2 = \frac{(2.819)^2(.04597)^2}{(.11847)^2(4.9321)} = 0.2426$

$d = \frac{(2.819)(.04597)}{.11847}\sqrt{\frac{24+1}{24}(1 - .242467) + \frac{(\hat{x}-2.40)^2}{4.9321}}$

$\hat{x}_L = 2.40 + \frac{1}{(1-.2426)}(\hat{x} - 2.40 - d)$

$\hat{x}_U = 2.40 + \frac{1}{(1-.2426)}(\hat{x} - 2.40 + d) \Rightarrow$

y	TRANS(y)	$L\hat{O}G(x)$	$\hat{x}$	d	$L\hat{O}G(x)_L$	$L\hat{O}G(x)_U$	$\hat{x}_L$	$\hat{x}_U$
10	.322	1.764	5.84	.242289	1.24038	1.88018	3.46	6.55
14	.383	2.285	9.83	.198634	1.98616	2.51068	7.29	12.31
19	.451	2.855	17.38	.221359	2.70876	3.29328	15.01	26.93

11.58 The four values of CUMVOL are 10, 20, 30, 12 yielding $\bar{y} = 0.42969$ and $s_{\bar{y}} = .05849$, where y is the arcsine of the square root of CUMVOL.

For 50%: $\hat{x} = 2.367$ and 95% confidence limits are (0.79, 4.45)

For 75%: $\hat{x} = 5.861$ and 95% confidence limits are (2.20, 12.88)

11.7 Correlation

11.59 The output yields R-square=0.9452. The estimated slope of the regression line is 0.0111049 which is positive, indicating an increasing relation between Branches and Business. Thus, the correlation is the positive square root of 0.9452, i.e., $r = 0.9722$.

11.60   a. Test $H_o : \rho_{yx} \leq 0$ versus $H_a : \rho_{yx} > 0$

Test Statistic: $t = \frac{.9722\sqrt{12-2}}{\sqrt{1-(.9722)^2}} = 13.13$

$p - value = Pr(t_{10} \geq 13.13) < 0.0005$

Conclusion: There is significant evidence that the correlation is positive.

   b. The t-test for testing whether the slope is 0 is exactly the same (except for rounding error) as the t-test for whether the correlation is 0. From the output we have $t = 13.138$, which is approximately the same as we computed in part a. In fact, they would be identical if we would have used a greater number of decimal places in our calculations in part a.

11.61   a. $r_{yx}^2 = 99.64\%$ (R-squared on output). This large value for r_{yx}^2 is reflected on the output by having the sum of squares for Model considerably larger than the sum of squares for Error. That is, virtually all the variation in TotalCost is accounted for by the variation in Runsize.

   b. Since β_1 is positive, there must be a general positive relation between the values for y and x. Thus, r_{yx} should be positive.

   c. In general, if the relation between the values of y and x are fairly consistent across the range of values for x, then a wider range of values for x yields a larger value for r_{yx}. Examining the plot in Exercise 11.39 confirms the consistency of the relation. Thus, we would expect r_{yx} to be smaller for the restricted data set.

11.63   a. There appears to be a general increase in salary as the level of experience increases. However, there is considerable variability in salary for persons having similar levels of experience.

   b. Note that Case 11 has a person with 14 years of experience and a salary of 37.9. Salaries of less than 40 are generally associated with persons having less than 5 years experience.

11.65 a. $\hat{y} = 40.507 + 1.470x$, where y is the starting salary and x is the years of experience. The slope value of 1.470 thousand dollars per years of experience can be interpreted in the following manner. A population of graduates with one extra year of experience would have an estimated average starting salary $1,470 higher than a population with one less year of experience. The intercept is the estimated average salary of graduates with no prior experience. Since the data included people with no prior experience, the estimated average salary for this group would in fact be the estimated intercept value of $40,507. (In situations where the data set does not contain points near x=0, the estimated intercept is not meaningful.)

b. The residual standard deviation is $s_\epsilon = 5.402$. This is an indication of the amount of variation in starting salaries which is not accounted for by the linear model relating starting salary to years of experience. If there are other factors affecting the variation in starting salary and/or if the relationship between starting salary and years of experience is nonlinear then s_ϵ would tend to be large.

c. From the output $t = 6.916$ with a p-value of 0.000. This would imply overwhelming evidence that there is a relation between starting salary and experience.

d. $R^2 = 0.494$ which indicates that 49.4% of the variation in starting salaries is accounted for by its linear relation with experience.

Supplementary Exercises

11.67 a. Scatterplot of the data is given here:

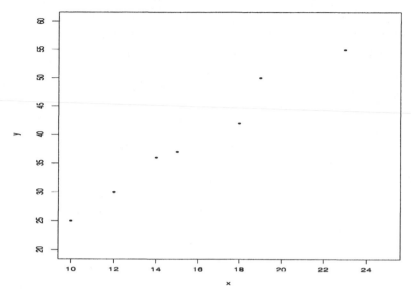

Plot of Data for Exercise 11.67

b. The necessary calculations are given here:

x	y	$(x - \bar{x})$	$(y - \bar{y})$	$(x - \bar{x})(y - \bar{y})$	$(x - \bar{x})^2$
10	25	-5.8571	-14.2857	83.6735	34.3061
12	30	-3.8571	-9.2857	35.8163	14.8776
14	36	-1.8571	-3.2857	6.1020	3.4490
15	37	-0.8571	-2.2857	1.9592	0.7347
18	42	2.1429	2.7143	5.8163	4.5918
19	50	3.1429	10.7143	33.6735	9.8776
23	55	7.1429	15.7143	112.2449	51.0204
111	275	0	0	279.2857	118.8571

$\bar{x} = \frac{111}{7} = 15.8571$ $\bar{y} = \frac{275}{7} = 39.2857$ $\hat{\beta}_1 = \frac{279.2857}{118.8571} = 2.3498 \Rightarrow$
$\beta_0 = 39.2857 - (2.3498)(15.8571) = 2.0247 \Rightarrow \hat{y} = 2.2047 + 2.3498x$

c. When $x_{n+1} = 21, \hat{y} = 2.2047 + (2.3498)(21) = 51.55$

11.68   a. The necessary calculations are given here:

x	y	$\hat{y}$	$(y - \hat{y})$	$(y - \hat{y})^2$
10	25	25.5228	-0.5228	0.2734
12	30	30.2224	-0.2224	0.0494
14	36	34.9219	1.0781	1.1624
15	37	37.2716	-0.2716	0.0738
18	42	44.3209	-2.3209	5.3866
19	50	46.6707	3.3293	11.0844
23	55	56.0697	-1.0697	1.1443
111	275	275	0	19.1743

101

b. $s_\epsilon = \sqrt{\frac{19.1743}{7-2}} = 1.9583$

All the residuals fall within $0 \pm 2s_\epsilon = (-3.9166, 3.9166)$

11.77   a. $t = 2.14$, with a p-value of 0.0396. With $\alpha = 0.05$, there is significant evidence of a linear relationship between rooms occupied and restaurant/lounge sales.

        b. From the scatterplot, it would appear that the point in the upper left corner of the plot is pulling the regression line towards the point. Removing this point, the regression line would undertake a counter clockwise twist which would result in an increase in its slope.

11.79   a. $\hat{y} = 9.7709 + 0.0501x$ and $s_\epsilon = 2.2021$

        b. 90% C.I. on β_1: $\hat{\beta}_1 \pm t_{.05,46} SE(\hat{\beta}_1) \Rightarrow 0.0501 \pm (1.679)(0.0009) \Rightarrow (0.0486, 0.0516)$

11.80 An examination of the data in the scatterplot indicates that two of the points may possibly be outliers since they are somewhat below the general pattern in the data. This may indicate that the data is nonnormal. Also, there appears to an increase in the variability of the Rate values as the Mileage increases. This would indicate that the condition of constant variance may be violated.

11.81 There is an error in the Excel output given in the book. Output from Minitab is given here:

Regression Analysis: Rate: versus Mileage:

The regression equation is
Rate: = 9.79 + 0.0503 Mileage:

```
Predictor      Coef      SE Coef        T        P
Constant      9.7932      0.4391      22.30    0.000
Mileage:    0.0503229   0.0008230     61.14    0.000
```

S = 2.042   R-Sq = 98.8%   R-Sq(adj) = 98.8%

Analysis of Variance

```
Source          DF        SS         MS        F        P
Regression       1       15590      15590    3738.63   0.000
Residual Error  46         192          4
Total           47       15782
```


Predicted Values for New Observations

```
New Obs    Fit     SE Fit       95.0% CI            95.0% PI
1        26.903     0.298   ( 26.303,  27.503)  ( 22.749,  31.057)
```

Values of Predictors for New Observations

```
New Obs  Mileage:
1          340
```

$$\hat{y} = 9.7932 + (0.0503229)(340) = 26.903$$

The 95% P.I. when x=340 is (22.749, 31.057)

11.87 The Minitab output is given here:

Regression Analysis: Yield versus Nitrogen

The regression equation is
Yield = 11.8 + 6.50 Nitrogen

```
Predictor      Coef      SE Coef        T        P
Constant      11.778      1.887        6.24     0.000
Nitrogen      6.5000      0.8736       7.44     0.000
```

S = 2.140   R-Sq = 88.8%   R-Sq(adj) = 87.2%

Analysis of Variance

```
Source          DF        SS         MS        F        P
Regression       1       253.50     253.50    55.36    0.000
Residual Error   7        32.06       4.58
  Lack of Fit    1         2.72       2.72     0.56     0.484
  Pure Error     6        29.33       4.89
Total            8       285.56
```

103

$\hat{y} = 11.778 + 6.5x$

From the output, $MS_{Lack} = 2.72$, $MS_{exp} = 4.89 \Rightarrow F = \frac{2.72}{4.89} = 0.56$ with df= 1, 6. This yields $p-value = 0.484$ which implies there is no indication of lack of fit in the model.

11.89   a. Scatterplot of the data is given here:

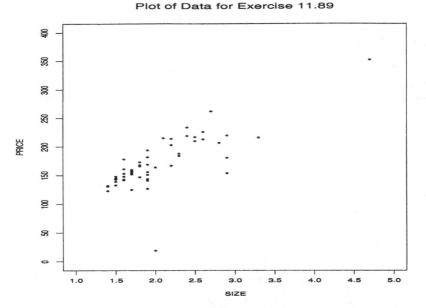

Plot of Data for Exercise 11.89

b. One of the houses in the study is shown as having a size of 2, but a very low price. It appears to be an outlier, with a value well below any reasonable line that may be fitted to the data. Because the point has an x-value near $\bar{x}$, the point does not have high leverage.

c. The minitab output is given here:

```
Regression Analysis: PRICE versus SIZE


The regression equation is
PRICE = 51.1 + 59.2 SIZE

Predictor        Coef      SE Coef         T        P
Constant        51.08        13.97      3.66    0.001
SIZE           59.152        6.666      8.87    0.000

S = 29.14              R-Sq = 58.9%        R-Sq(adj) = 58.1%
PRESS = 50352.0        R-Sq(pred) = 55.66%

Analysis of Variance

Source            DF          SS        MS        F        P
Regression         1       66862     66862    78.73    0.000
Residual Error    55       46707       849
Total             56      113569

Unusual Observations
Obs     SIZE     PRICE       Fit     SE Fit    Residual    St Resid
  4     4.70    352.00    329.09      18.32       22.91       1.01 X
 27     2.00     19.10    169.38       3.86     -150.28      -5.20R
 47     2.90    154.00    222.62       7.06      -68.62      -2.43R

R denotes an observation with a large standardized residual
X denotes an observation whose X value gives it large influence.

Durbin-Watson statistic = 2.10
```

The least squares line is $\hat{y} = 51.08 + 59.152x$

d. Minitab output for the data set without the outlier is given here:

```
Regression Analysis: PRICE versus SIZE


The regression equation is
PRICE = 54.0 + 59.0 SIZE

Predictor        Coef      SE Coef         T        P
Constant        53.99        10.06      5.37    0.000
SIZE           59.040        4.794     12.31    0.000

S = 20.96              R-Sq = 73.7%        R-Sq(adj) = 73.3%
PRESS = 26440.0        R-Sq(pred) = 70.73%

Analysis of Variance

Source            DF          SS        MS        F        P
Regression         1       66607     66607   151.64    0.000
Residual Error    54       23719       439
Total             55       90327

Unusual Observations
Obs     SIZE     PRICE       Fit     SE Fit    Residual    St Resid
  4     4.70    352.00    331.48      13.18       20.52       1.26 X
```

13	2.90	181.00	225.20	5.09	-44.20	-2.17R
26	2.70	262.00	213.40	4.32	48.60	2.37R
46	2.90	154.00	225.20	5.09	-71.20	-3.50R

```
R denotes an observation with a large standardized residual
X denotes an observation whose X value gives it large influence.

Durbin-Watson statistic = 1.83

Predicted Values for New Observations

New Obs    Fit    SE Fit      95.0% CI           95.0% PI
1        349.19    14.59  ( 319.94,  378.43) ( 297.99,  400.38) XX
X  denotes a row with X values away from the center
XX denotes a row with very extreme X values

Values of Predictors for New Observations

New Obs    SIZE
1          5.00
```

The least squares line is $\hat{y} = 53.99 + 59.04x$. Note that the slope has changed very little (59.152 versus 59.04) since the outlier was of low leverage.

e. The residual standard deviations are $s_\epsilon = 29.14$ with outlier and $s_\epsilon = 20.96$ without outlier. There has been a considerable reduction in the residual standard deviation because its the squared distance from the fitted line and hence is very sensitive to outliers.

11.90 a. The estimated intercept is $\hat{\beta}_o = 53.99$. This is the estimated mean price of houses of size 0. This could be interpreted as the estimated price of land upon which there is no building. However, there were no data values with x near 0. Therefore, the estimated intercept should not be directly interpreted but just taken as a portion of an overall model.

b. A slope of 0 would indicate that the estimated mean price of houses does not increase as the size of the house increases. That is, large houses have the same price as small houses. This is not very realistic. From the Minitab output, $t = 12.31$ with df=54, $\Rightarrow p - value = Pr(t_{54} \geq 12.31) < 0.0005$. Thus, there is highly significant evidence evidence that the slope is not 0.

c. Using the estimates from the Minitab output, a 95% C.I. for β_1 is $59.040 \pm (2.005)(4.794) \Rightarrow$ $(49.428, 68.652)$.

11.91 a. From the Minitab output, a 95% P.I. for the selling price when the size is 5: $(297.99, 400.38)$ The prediction may be somewhat suspect since a house of size 5 (5000 square feet) is beyond all the data values in the study.

b. Scatterplot of the data is given here:

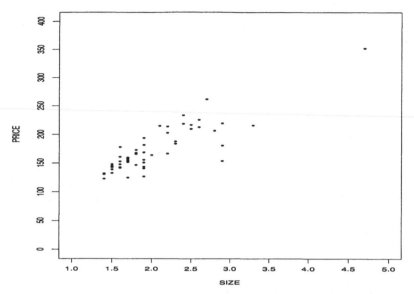

Plot of Data for Exercise 11.91

The variance in selling price appears to increase in variability as the size of the houses increases. For small houses, the selling prices are concentrated near the fitted line. For large houses, there is a wide spread in the selling price for houses of nearly the same size. Therefore, the assumption of constant variance may not be valid.

    c. The P.I. may not be valid since the statistical procedures depend on the condition that the variance remain constant across the range of x-values.

11.96 Minitab output is given here:

```
Regression Analysis: DURABIL versus CONCENTR


The regression equation is
DURABIL = 47.0 + 0.308 CONCENTR

Predictor       Coef      SE Coef        T        P
Constant      47.020        4.728     9.94    0.000
CONCENTR      0.3075       0.1114     2.76    0.008

S = 12.21              R-Sq = 11.6%        R-Sq(adj) = 10.1%
PRESS = 9368.62        R-Sq(pred) = 4.20%

Analysis of Variance

Source          DF          SS         MS        F        P
Regression       1      1134.7     1134.7     7.61    0.008
Residual Error  58      8644.5      149.0
Total           59      9779.2

Unusual Observations
```

Obs	CONCENTR	DURABIL	Fit	SE Fit	Residual	St Resid
2	20.0	25.20	53.17	2.73	-27.97	-2.35R
4	20.0	20.30	53.17	2.73	-32.87	-2.76R
60	60.0	18.90	65.47	2.73	-46.57	-3.91R

R denotes an observation with a large standardized residual

Durbin-Watson statistic = 1.35

    a. $\hat{y} = 47.020 + 0.3075x$. The estimated slope $\hat{\beta}_1 = 0.3075$ can be interpreted as follows: there is a 0.3075 increase in average durability for a 1 unit increase in the concentration.

    b. The coefficient of determination, $R^2 = 11.6\%$. That is, 11.6% of the variation in durability is explained by its linear relationship with concentration. Thus, a straight-line model relating Durability to Concentration would not yield very accurate predictions.

11.97 For testing $H_o : \beta_1 = 0$ versus $H_a : \beta_1 \neq 0$, $t = 2.76$ with $p-value = 0.008$. The p-value is less than $\alpha = 0.01$ thus we reject H_o and conclude there is significant evidence that the slope is different from 0.

11.98 Scatterplot of the data is given here:

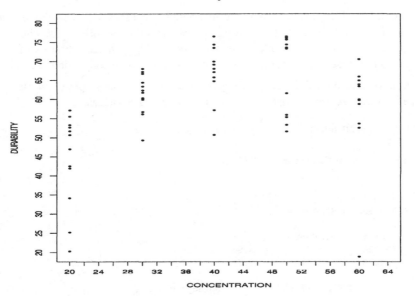

Plot of Durability vs Concentration

    a. From the scatterplot, there is a definite curvature in the relation between Durability and Concentration. A straight-line model would not appear to be appropriate.

    b. The coefficient of determination, R^2, measures the strength of the linear (straight-line) relation only. A straight-line model does not adequately describe the relation between Durability and Concentrtation. This is indicated by the small percentage of the variation, 11.6%, in the values of Durability explained by the model containing just a linear relation with Concentration. A more complex relation exists between the variables Durability and Concentration.

Chapter 12: Multiple Regression and the General Linear Model

12.2 The General Linear Model

12.1 a. Suppose the three qualitative variables, V_1, V_2, V_3, have two, three and two levels, respectively and we observe the response, y during n experiments involving these variables. In order to model this experiment, it is necessary to create the $(2)(3)(2)=12$ combinations consisting one level from each of the three variables and denote their means by μ_i, with $i = 1, 2, \cdots, 12$. For the jth running of the experiment, $j = 1, 2, \cdots, n$, the model is

$$y_j = \mu_{i_j} + \epsilon_j, \text{ with } i_j = 1, 2, \cdots, 12 \text{ representing which one of the 12 combinations}$$
of the qualitative variables that was observed with y_j.

b. To write the above model in a general linear model form, we have to create dummy variables to indicate which levels of the three qualitiative variables were observed along with the dependent variable y during the n runnings of the experiment:

$$x_1 = \begin{cases} 1 & \text{if } V_1 = \text{1st Level} \\ 0 & \text{if } \text{Otherwise} \end{cases} \qquad x_2 = \begin{cases} 1 & \text{if } V_2 = \text{1st Level} \\ 0 & \text{if } \text{Otherwise} \end{cases}$$

$$x_3 = \begin{cases} 1 & \text{if } V_2 = \text{2nd Level} \\ 0 & \text{if } \text{Otherwise} \end{cases} \qquad x_4 = \begin{cases} 1 & \text{if } V_3 = \text{1st Level} \\ 0 & \text{if } \text{Otherwise} \end{cases}$$

$$y_j = \beta_o + \beta_1 x_{1j} + \beta_2 x_{2j} + \beta_3 x_{3j} + \beta_4 x_{4j} + \epsilon_j, \text{ with } j = 1, \cdots, n$$

12.3 The model is given here:

$$y_i = \beta_0 + \beta_1 x_{1i} + \beta_2 x_{2i} + \beta_3 x_{3i} + \beta_4 x_{4i} + \beta_5 x_{5i} + \beta_6 x_{6i} + \beta_7 x_{7i} + \beta_8 x_{8i} + \beta_9 x_{9i} + \epsilon_i$$
with $x_1 = x_1$, $x_2 = x_2$, $x_3 = x_3$, $x_4 = x_1^2$, $x_5 = x_2^2$,
$x_6 = x_3^2$, $x_7 = x_1 x_2$, $x_8 = x_1 x_3$, $x_9 = x_2 x_3$

12.5 a. For the model of the ith observation in the experiment:

$$y_i = \beta_o + \beta_1 x_{1i} + \beta_2 x_{2i} + \beta_3 x_{3i} + \epsilon_i,$$

an identification of the parameters is given here with the notation μ_{jk} being the mean of the population consisting of Treatment #j and Location #k; $j = 1, 2, 3$; $k = 1, 2$:

	Location	
Treatment	1	2
1	$\mu_{11} = \beta_o$	$\mu_{12} = \beta_o + \beta_3$
2	$\mu_{21} = \beta_o + \beta_1$	$\mu_{22} = \beta_o + \beta_1 + \beta_3$
3	$\mu_{31} = \beta_o + \beta_2$	$\mu_{32} = \beta_o + \beta_2 + \beta_3$

We can thus interpret the $\beta's$ as follows:

$\beta_o = \mu_{11}$, mean of TRT #1, LOC #1

$\beta_1 = \mu_{21} - \mu_{11}$ or $\beta_1 = \mu_{22} - \mu_{12}$, difference of mean of TRT #2 and TRT #1 at a given Location

$\beta_2 = \mu_{31} - \mu_{11}$ or $\beta_2 = \mu_{32} - \mu_{12}$,

difference of mean of TRT #3 and TRT #1 at a given Location

$\beta_3 = \mu_{12} - \mu_{11}$ or $\beta_3 = \mu_{22} - \mu_{21}$ or $\beta_3 = \mu_{32} - \mu_{31}$,

difference of mean of LOC #2 and LOC #1 for a given Treatment

b. $\mu_{22} - \mu_{32} = (\beta_0 + \beta_1 + \beta_3) - (\beta_0 + \beta_2 + \beta_3) = \beta_1 - \beta_2$

Yes, the difference is the same for Location #1.

12.3 Estimating Multiple Regression Coefficients

12.7   a. A scatterplot of the data is given here:

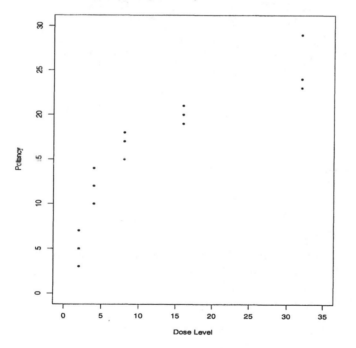

b. $\hat{y} = 8.667 + 0.575x$

c. From the scatterplot, there appears to be curvature in the relation between Potency and Dose Level. A quadratic or cubic model may provide an improved fit to the data.

d. The quadratic model provides the better fit. The quadratic model has a much lower MS(Error), its R^2 value is 11% larger, the quadratic term has a p-value of 0.0062 which indicates that this term is significantly different from 0, however, the residual plot still has a distinct curvature as was found in the residual plot for the linear model.

12.8 a. The logarithm of the dose levels are given here:

Dose Level (x)	2	4	8	16	32
ln(x)	0.693	1.386	2.079	2.773	3.466

A scatterplot of the data is given here:

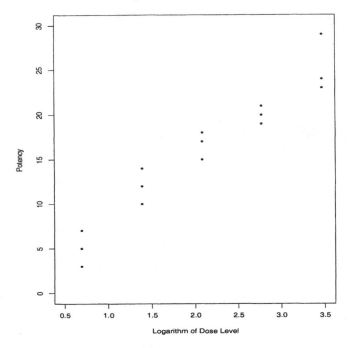

b. $\hat{y} = 1.2 + 7.021 ln(x)$

c. The model using $ln(x)$ provides a better fit based on the scatterplot, the decrease in MS(Error) over the quadratic model, increase in R^2, and the residual plot appears to be a random scatter of points about the horizontal line, whereas there was a curvature in the residual plot from the fit of the quadratic model.

12.9 a. $\hat{y} = 326.39 + 136.10 * Promo - 61.18 * Devel - 43.70 * Research$

b. $s_\epsilon = \sqrt{MS(Residual)} = \sqrt{656.811614} = 25.628$

c. For fixed values of Devel and Research, a 1 unit increase in Promo yields a 136.10 increase in the average value of Sales.

12.4 Inferences in Multiple Regression

12.17   a. MS(Regression) = 159.67 and MS(Residual) = 7.00

   b. F = 22.81

   c. $p - value = Pr(F_{3,8} \geq 22.81) < 0.0001$

   d. In testing $H_o : \beta_1 = \beta_2 = \beta_3 = \beta_4 = 0$ versus H_a : at least one $\beta_i \neq 0$, the p-value provides strong evidence to reject H_o and conclude that the independent variables x,w, and v, as a group, have at least some degree of relationship with the dependent variable, y.

   e. A 95% C.I. for β_1 is given by $\hat{\beta}_1 \pm (t_{.025,8})(SE(\hat{\beta}_1)) \Rightarrow 5.000 \pm (2.306)(6.895) \Rightarrow (-10.90, 20.90)$

   This is very wide interval and the interval includes 0. Thus, we can conclude the variable x adds very little predictive power to a model which already contains the variables w and v.

12.19   To assess whether each of the variables is contributing to the predictive value of the model, we will examine the t-statistics for each independent variable, separately. For Air Miles, t=2.43 with p-value=0.0253. Therefore, there is evidence that Air Miles adds predictive value to the model beyond that provided by Population. For Population, t=8.80 with $p - value \approx 0.0000$. Therefore, there is evidence that Population adds substantial predictive value to the model beyond that provided by Air Miles.

12.20   The t-value is $t_{.05,19} = 1.729 \Rightarrow$ 90% C.I.'s for β_i are given by:

   Air Miles: $0.2922116 \pm (1.729)(0.120336) \Rightarrow (0.0841, 0.5003)$

   Population: $1.5310653 \pm (1.729)(0.174004) \Rightarrow (1.2302, 1.8319)$

12.21   a. $\hat{y}$ 7.20430 + 1.36291 METAL+0.30588 TEMP+0.01024 WATTS−0.00277 METXTEXP

   b. The results of the various t-tests are given here:

H_o	H_a	T.S. t	Conclusion
$\beta_o = 0$	$\beta_o \neq 0$	$t = 0.41$	$p - value = 0.6855$ Fail to Reject H_o
$\beta_1 = 0$	$\beta_1 \neq 0$	$t = 1.47$	$p - value = 0.1559$ Fail to Reject H_o
$\beta_2 = 0$	$\beta_2 \neq 0$	$t = 0.19$	$p - value = 0.8522$ Fail to Reject H_o
$\beta_3 = 0$	$\beta_3 \neq 0$	$t = 2.16$	$p - value = 0.0427$ Reject H_o
$\beta_4 = 0$	$\beta_4 \neq 0$	$t = -0.04$	$p - value = 0.9717$ Fail to Reject H_o

   Of the four independent variables, only WATTS appears to have predictive value given the remaining three variables have already been included in the model.

   c. $t_{.025,20} = 2.086 \Rightarrow$ 95% C.I. on β_4 is given by

   $-0.00277 \pm (2.086)(0.07722) \Rightarrow (-0.164, 0.158)$

   d. VIF measures how much the standand error of a regression coefficient (β_i) is increased due to collinearity. If the value of VIF is very large, such as 10 or more, collinearity is a serious problem. The variables TEMP and METXTEMP have VIF values

113

extremely large (250 and 246.4, respectively). An examination of the Pearson Correlations reveals that the correlation between TEMP and METXTEMP is 0.9831, that is, nearly a perfect correlation between the two variables. One of the variables, TEMP or METXTEMP, should be removed from the model and the coefficients of the remaining variables recomputed.

12.5 Testing a Subset of Regression Coefficients

12.23   a. $R^2 = 0.6978$

     b. In the complete model, we want to test $H_o : \beta_2 = \beta_3 = 0$ versus $H_a : \beta_2 \neq 0$ and/or $\beta_3 \neq 0$. The F-statistic has the form:
$$F = \frac{[SSReg.,Complete - SSReg.,Reduced]/(k-g)}{SSResidual,Complete/[n-(k+1)]} = \frac{[43901.7677 - 39800.7248]/(3-1)}{13136.2323/[24-4]} = 3.12$$
with $df = 2, 20 \Rightarrow p-value = Pr(F_{2,20} \geq 3.12) = 0.066 \Rightarrow$

Fail to reject H_o. There is not substantial evidence to conclude that either $\beta_2 \neq 0$ or $\beta_3 \neq 0$.

     c. Based on the F-test, omitting Devel and Research from the model would not substantially change the fit of the model. Neither Devel nor Research appear to add any predictive value to the model containing Promo.

12.25 In the complete model, we want to test $H_o : \beta_1 = \beta_2 = 0$ versus $H_a : \beta_1 \neq 0$ and/or $\beta_2 \neq 0$. The F-statistic has the form:
$$F = \frac{[SSReg.,Complete - SSReg.,Reduced]/(k-g)}{SSResidual,Complete/[n-(k+1)]} = \frac{[2.65376 - 0.68192]/(3-1)}{0.67461/[21-4]} = 24.84$$
with $df = 2, 17 \Rightarrow p-value = Pr(F_{2,17} \geq 24.84) < 0.0001 \Rightarrow$

Reject H_o. There is substantial evidence to conclude that $\beta_2 \neq 0$ and/or $\beta_3 \neq 0$. Based on the F-test, omitting BUSIN and COMPET from the model has substantially changed the fit of the model. Dropping one or both of these independent variables from the model will result in a decrease in the predictive value of the model.

12.28   a. $\hat{y} = 50.0195 + 6.64357x_1 + 7.3145x_2 - 1.23143x_1^2 - 0.7724x_1x_2 - 1.1755x_2^2$

     b. $\hat{y} = 70.31 - 2.676x_1 - 0.8802x_2$

     c. For the Complete model: $R^2 = 86.24\%$
       For the Reduced model: $R^2 = 58.85\%$

     d. In the complete model, we want to test

       $H_o : \beta_3 = \beta_4 = \beta_5 = 0$ versus $H_a :$ at least one of $\beta_3, \beta_4, \beta_5 \neq 0$.

       The F-statistic has the form:
$$F = \frac{[SSReg.,Complete - SSReg.,Reduced]/(k-g)}{SSResidual,Complete/[n-(k+1)]} = \frac{[448.193 - 305.808]/(5-2)}{71.489/[20-6]} = 9.29$$
       with $df = 3, 14 \Rightarrow p-value = Pr(F_{3,14} \geq 9.29) = 0.0012 \Rightarrow$

Reject H_o. There is substantial evidence to conclude that at least one of $\beta_3, \beta_4, \beta_5 \neq 0$. Based on the F-test, omitting the second order terms from the model has substantially changed the fit of the model. Dropping one or more of these independent variables from the model will result in a decrease in the predictive value of the model.

12.6 Forecasting Using Multiple Regression

12.29 The predicted y-value at x=3, w=1, v=6 is $\hat{y} = 33.000$ with 95% P.I.: (21.788, 44.212). The selected values of the independent variables are at the extremes of the data used to fit the model. Therefore, the prediction is being identified as being computed at "very extreme X values".

12.30   a. For the second order model, the 95% P.I. are
$\quad\quad$ (54.7081, 65.1439) for $x_1 = 3.5$ and $x_2 = 0.35$
$\quad\quad$ (57.0829, 67.6529) for $x_1 = 3.5$ and $x_2 = 2.5$

$\quad$ b. For the first order model, the 95% P.I. are
$\quad\quad$ (50.0280, 65.6986) for $x_1 = 3.5$ and $x_2 = 0.35$
$\quad\quad$ (51.0525, 66.4345) for $x_1 = 3.5$ and $x_2 = 2.5$

$\quad$ c. The width of the P.I.'s from the second order model are less than the width of the P.I.'s from the first order model. The prediction limits are tighter for the second order model since it provides a better fitting model and hence has smaller standard deviations for the predictions.

12.7 Comparing the Slopes of Several Regression Lines

12.31   a. $y = \beta_o + \beta_1 x_1 + \beta_2 x_2 + \beta_3 x_3 + \beta_4 x_1 x_2 + \beta_5 x_1 x_3 + \epsilon$, where
$\quad\quad x_1 = \log(\text{dose})$

$$x_2 = \begin{cases} 1 & \text{if} \quad \text{Product B} \\ 0 & \text{if} \quad \text{Products A or C} \end{cases} \quad\quad x_3 = \begin{cases} 1 & \text{if} \quad \text{Product C} \\ 0 & \text{if} \quad \text{Products A or B} \end{cases}$$

$\quad\quad \beta_0 = y-$intercept for Product A regression line
$\quad\quad \beta_1 = $ slope for Product A regression line
$\quad\quad \beta_2 = $ difference in y-intercepts for Products A and B regression lines
$\quad\quad \beta_3 = $ difference in y-intercepts for Products A and C regression lines
$\quad\quad \beta_4 = $ difference in slopes for Products A and B regression lines
$\quad\quad \beta_5 = $ difference in slopes for Products A and C regression lines

$\quad$ b. $y = \beta_o + \beta_1 x_1 + \beta_2 x_1 x_2 + \beta_3 x_1 x_3 + \epsilon$

12.32   a. The Minitab output for fitting the Complete and Reduced models is given here:

Regression Analysis: y versus x1, x2, x3, x1*x2, x1*x3

The regression equation is
y = 7.31 + 3.30 x1 - 2.15 x2 - 4.35 x3 - 1.50 x1*x2 - 2.28 x1*x3

Predictor	Coef	SE Coef	T	P
Constant	7.3072	0.2103	34.75	0.000
x1	3.3038	0.2186	15.11	0.000
x2	-2.1548	0.2974	-7.25	0.000
x3	-4.3486	0.2974	-14.62	0.000
x1*x2	-1.5004	0.3092	-4.85	0.003
x1*x3	-2.2795	0.3092	-7.37	0.000

S = 0.3389   R-Sq = 98.8%   R-Sq(adj) = 97.7%

Analysis of Variance

Source	DF	SS	MS	F	P
Regression	5	55.293	11.059	96.30	0.000
Residual Error	6	0.689	0.115		
Total	11	55.982			

Regression Analysis: y versus x1, x2, x3

The regression equation is
y = 6.59 + 2.04 x1 - 1.30 x2 - 3.05 x3

Predictor	Coef	SE Coef	T	P
Constant	6.5894	0.5131	12.84	0.000
x1	2.0438	0.3519	5.81	0.000
x2	-1.3000	0.6679	-1.95	0.087
x3	-3.0500	0.6679	-4.57	0.002

S = 0.9446   R-Sq = 87.2%   R-Sq(adj) = 82.5%

Analysis of Variance

Source	DF	SS	MS	F	P
Regression	3	48.844	16.281	18.25	0.001
Residual Error	8	7.138	0.892		
Total	11	55.982			

In the complete model: $y = \beta_o + \beta_1 x_1 + \beta_2 x_2 + \beta_3 x_3 + \beta_4 x_1 x_2 + \beta_5 x_1 x_3 + \epsilon$, the test of equal slopes is a test of the hypotheses:

$H_o : \beta_4 = 0, \beta_5 = 0$ versus $H_o : \beta_4 \neq 0$ and/or $\beta_5 \neq 0$

Under H_o, the reduced model becomes $y = \beta_o + \beta_1 x_1 + \beta_2 x_2 + \beta_3 x_3 + \epsilon$

$F = \frac{(55.293 - 48.844)/(5-3)}{0.689/6} = 28.08 \Rightarrow p - value = Pr(F_{2,6} \geq 28.08) = 0.0009 \Rightarrow$

b. Reject H_o and conclude there is significant evidence that the slopes of the three regression lines (one for each Drug Product) are different.

c. In the complete model, a test of equal intercepts is a test of the hypotheses:

$H_o : \beta_2 = 0, \beta_3 = 0$ versus $H_o : \beta_2 \neq 0$ and/or $\beta_3 \neq 0$

Under H_o, reduced model becomes $y = \beta_o + \beta_1 x_1 + \beta_4 x_1 x_2 + \beta_5 x_1 x_3 + \epsilon$

Obtain the SS's from the reduced model and then conduct the F-test as was done in part a.

12.8 Logistic Regression

12.33   a. For testing $H_o : \beta_1 = 0$ versus $H_a : \beta_1 \neq 0$, the p-value for the output is $p - value = 0.0427$. Thus, at the $\alpha = 0.05$ level we can reject H_o and conclude there is significant evidence that the probability of Completing the Task is related to the amount of Experience.

   b. From the output, $\hat{p}(24) = 0.765$ with 96% C.I. (0.437, 0.932).

Supplementary Exercises

12.35   a. $\hat{y} = -1.320 + 5.550$ EDUC$+0.885$ INCOME$+1.925$ POPN-11.389 FAMSIZE
           (57.98)   (2.702)   (1.308)   (1.371)   (6.669)

   b. $R^2 = 96.2\%$ and $s_\epsilon = 2.686$

   c. A value of 2.07 standard deviations from the predicted value is unusual since we would expect approximately 95% of all values to be within 2 standard deviations of the predicted. Thus, 2.07 standard deviations from the predicted is a moderately unusual point, but not a serious outlier.

12.37   a. $R^2 = 94.2\%$

   b. In the complete model, we want to test
   $H_o : \beta_2 = \beta_3 = 0$ versus $H_a :$ at least one of $\beta_2, \beta_3 \neq 0$.
   The F-statistic has the form:
   $F = \frac{[SSReg.,Complete - SSReg.,Reduced]/(k-g)}{SSResidual,Complete/[n-(k+1)]} = \frac{[1295.70 - 1268.48]/(4-2)}{50.51/[12-5]} = 1.89$
   with $df = 2, 7 \Rightarrow p - value = Pr(F_{2,7} \geq 1.89) = 0.2206 \Rightarrow$
   Fail to reject H_o. There is not substantial evidence to conclude that $\beta_2 \neq 0$ and/or $\beta_3 \neq 0$. Based on the F-test, omitting INCOME and POPN from the model would not substantially changed the fit of the model. Dropping these independent variables from the model will not result in a large decrease in the predictive value of the model.

12.39 a. $F = \dfrac{0.894477/4}{(1-0.894477)/(43-5)} = 80.53$ with df=4,38. The $p-value = Pr(F_{4,38} \geq 80.53) < 0.0001 \Rightarrow$ Reject $H_o : \beta_1 = \beta_2 = \beta_3 = \beta_4 = 0$ and conclude that at least one of the four independent variables has predictive value for Loan Volume.

b. Using $\alpha = 0.01$, none of the p-values for testing $H_o : \beta_i = 0$ versus $H_a : \beta_i \neq 0$, .0999, .0569, .5954, and .3648, respectively, are less than 0.01. Thus, none of the independent variables provide substantial predictive value given the remaining three variables in the model. That is, given a model with three variables included in the model, the fourth variable does not add much by including it also.

c. The contradiction is due to the severe collinearity that is present in the four independent variables. The F test demonstrates that as a group the four independent variables provide predictive value, but because the four independent variables are highly correlated the information concerning their relationship with the dependent variable, Loan Volume, is highly overlapping. Thus, it is very difficult to determine which of the independent variables are useful in predicting Loan Volume.

12.41 a. The regression model is

$\hat{y} = -16.8198 + 1.47019x_1 + .994778x_2 - .0240071x_3 - .01031x_4 - .000249574x_5$

$s_\epsilon = 3.39011$

b. Test $H_o : \beta_3 = 0$ versus $H_a : \beta_3 \neq 0$. From output, $t = -1.01$ with p-value=0.3243. Thus, there is not substantial evidence that the variable $x_3 = x_1x_2$ adds predictive value to a model which contains the other four independent variables.

12.43 a. $\hat{y} = 0.8727 + 2.548$ SIZE $+ 0.220$ PARKING $+ 0.589$ INCOME

   (1.946)　(1.201)　　(0.155)　　　　(0.178)

b. The interpretation of coefficients is given here:

Coefficient	Interpretation
$\hat{\beta}_0 = y-$Intercept	The estimated average daily sales for the population of stores having 0 Size, 0 Parking, 0 Income
$\hat{\beta}_1 = \hat{\beta}_{SIZE}$	The estimated change in average Daily Sales per unit change in SIZE, for fixed values of PARKING and INCOME
$\hat{\beta}_2 = \hat{\beta}_{PARKING}$	The estimated change in average Daily Sales per unit change in PARKING, for fixed values of SIZE and INCOME
$\hat{\beta}_3 = \hat{\beta}_{INCOME}$	The estimated change in average Daily Sales per unit change in INCOME, for fixed values of SIZE AND PARKING

c. $R^2 = 0.7912$ and $s_\epsilon = 0.7724$

d. Only the pairwise correlations between the independent variables is given on output. A better indicator of collinearity is the values for VIF or the R^2 values from predicting each independent variable from the remaining independent variables. Examining the correlations does not reveal any very large values. Only SIZE and PARKING with a correlation of 0.6565 appear to be near a value which would be of concern relative to collinearity.

12.44 The results of the various tests are given here:

H_o	H_a	T.S.	$p - value$	Conclusion
$\beta_1 = \beta_2 = \beta_3 = 0$	at least one $\neq 0$	$F = 15.16$	0.0002	Reject H_o
$\beta_o = 0$	$\beta_o \neq 0$	$t = 0.449$	0.662	Fail to Reject H_o
$\beta_1 = 0$	$\beta_1 \neq 0$	$t = 2.122$	0.055	Fail to Reject H_o
$\beta_2 = 0$	$\beta_2 \neq 0$	$t = 1.418$	0.182	Fail to Reject H_o
$\beta_3 = 0$	$\beta_3 \neq 0$	$t = 3.310$	0.006	Reject H_o

The significance of the F-test implies that the model with all three independent variables has some predictive value in predicting SALES. The t-tests indicate that only INCOME has additional predictive value given the other two independent variables are already in the model. SIZE and SALES do not appear to have additional predictive value when the model already contains the other two independent variables.

12.45 a. $\hat{y} = 102.708 - .833$ PROTEIN $- 4.000$ ANTIBIO $- 1.375$ SUPPLEM

b. $s_\epsilon = 1.70956$

c. $R^2 = 90.07\%$

d. There is no collinearity problem in the data set. The correlations between the pairs of independent variables is 0 for each pair and the VIF values are all equal to 1.0. This total lack of collinearity is due to the fact that the independent variables are perfectly balanced. Each combination of PROTEIN and ANTIBIO values appear exactly three times in the data set. Each combination of PROTEIN and SUPPLEM occur twice, etc.

12.46 a. When PROTEIN=15%, ANTIBIO=1.5%, SUPPLEM=5%,
$\hat{y} = 102.708 - .83333(15) - 4.000(1.5) - 1.375(5) = 77.333$

b. There is no extrapolation for these values of the independent variables because these values represent the mean of the values in the data set. In other words, the prediction of y is occurring in the middle of the data set.

c. The 95% C.I. on the mean value of TIME when PROTEIN=15%, ANTIBIO=1.5%, SUPPLEM=5% is given on the output: (76.469, 78.197)

12.47 a. $\hat{y} = 89.8333 - 0.83333$ PROTEIN

b. $R^2 = 0.5057$

c. In the complete model, we want to test
$H_o : \beta_2 = \beta_3 = 0$ versus $H_a :$ at least one of $\beta_2, \beta_3 \neq 0$.
The F-statistic has the form:
$F = \frac{[371.083 - 208.333]/(3-1)}{40.9166/[18-4]} = 27.84$
with $df = 2, 14 \Rightarrow p - value = Pr(F_{2,14} \geq 27.84) < 0.0001 \Rightarrow$ Reject H_o.
There is substantial evidence to conclude that at least one of $\beta_2, \beta_3 \neq 0$. Based on the F-test, omitting x_2 and/or x_3 from the model would substantially changed the fit of the model. Dropping ANTIBIO and/or SUPPLEM from the model may result in a large decrease in the predictive value of the model.

12.52 SAS output is given here:

The REG Procedure
Model: MODEL1
Dependent Variable: y SALARY

Analysis of Variance

Source	DF	Sum of Squares	Mean Square	F Value	Pr > F
Model	3	39.31705	13.10568	13.11	<.0001
Error	63	62.95698	0.99932		
Corrected Total	66	102.27403			

Root MSE	0.99966	R-Square	0.3844	
Dependent Mean	29.36418	Adj R-Sq	0.3551	
Coeff Var	3.40435			

Parameter Estimates

Variable	Label	DF	Parameter Estimate	Standard Error	t Value	Pr > |t|
Intercept	Intercept	1	25.53758	0.64279	39.73	<.0001
x1	NUMEXPL	1	0.00389	0.00172	2.27	0.0269
x2	MARGIN	1	0.09567	0.03652	2.62	0.0110
x3	IPCOST	1	0.21657	0.06917	3.13	0.0026

a. $\hat{y} = 25.53758 + 0.00389$ NUMEXPL $+ 0.09567$ MARGIN $+ 0.21657$ IPCOST

The coefficient for NUMEXPL indicates that, for firms with equal MARGIN and IP-COST, a firm with one more employee is estimated to pay an extra 0.00389 (thousand dollars) in average salary. The coefficient for MARGIN indicates that, for firms with equal NUMEXPL and IPCOST, a firm with one-unit higher MARGIN is estimated to pay an extra 0.09567 (thousand dollars) in average salary. The coefficient for IPCOST indicates that, for firms with equal NUMEXPL and MARGIN, a firm with one-unit higher IPCOST is estimated to pay an extra 0.21657 (thousand dollars) in average salary.

b. The F-statistic from the AOV on the output is $F = 13.11$ with $p-value < 0.0001$. There is highly significant evidence that the independent variables as a group provide some predictive value for estimating salary.

c. Using the $p-values$ from the output, all three independent variables have relatively small p-values (0.0269, 0.0110, 0.0026). Thus, each of the three independent variables provide added predictive value to a model containing only two of the independent variables.

12.53   a. $R^2 = 0.3844 = 38.44\%$

b. The SAS output is given here:

The SAS System

Dependent Variable: y SALARY

Analysis of Variance

Source	DF	Sum of Squares	Mean Square	F Value	Pr > F
Model	1	3.66167	3.66167	2.41	0.1251
Error	65	98.61236	1.51711		
Corrected Total	66	102.27403			

Root MSE	1.23171	R-Square	0.0358	
Dependent Mean	29.36418	Adj R-Sq	0.0210	
Coeff Var	4.19461			

Parameter Estimates

Variable	Label	DF	Parameter Estimate	Standard Error	t Value	Pr > |t|
Intercept	Intercept	1	29.08407	0.23484	123.85	<.0001
x1	NUMEXPL	1	0.00326	0.00210	1.55	0.1251

R^2 has decreased dramatically to 0.0358=3.58%.

c. In the complete model, we want to test

$H_o : \beta_2 = \beta_3 = 0$ versus H_a : at least one of $\beta_2, \beta_3 \neq 0$.

The F-statistic has the form:

$F = \frac{[39.31706 - 3.66167]/(3-1)}{62.95698/[67-4]} = 17.84$

with $df = 2, 63 \Rightarrow p - value = Pr(F_{2,63} \geq 17.84) < 0.0001 \Rightarrow$ Reject H_o.

There is substantial evidence to conclude that at least one of $\beta_2, \beta_3 \neq 0$. Based on the F-test, omitting MARGIN and/or IPCOST from the model would substantially changed the fit of the model. Dropping MARGIN and IPCOST from the model will result in a large decrease in the predictive value of the model.

12.54 SAS output is given here:

The SAS System

Pearson Correlation Coefficients, N = 67

	SALARY	NUMEXPL	MARGIN	IPCOST
SALARY	1.00000	0.18922	0.50447	0.50727
NUMEXPL	0.18922	1.00000	0.00884	-0.10725
MARGIN	0.50447	0.00884	1.00000	0.53134
IPCOST	0.50727	-0.10725	0.53134	1.00000

Only the correlation between MARGIN and IPCOST, 0.53134, is of a moderate size. The other two correlations among the independent variables is very small in magnitude (.10725 and .00884). Therefore, collinearity does not appear to be a problem.

12.58   a. Holding any three of the four predictor variables constant, we would expect that a unit increase in the remaining variable would result in an increase in sales. Thus, we would expect that each of the four partial slopes would be positive.

b. SAS output is given here:

The REG Procedure
Model: MODEL1
Dependent Variable: y SALES

Analysis of Variance

Source	DF	Sum of Squares	Mean Square	F Value	Pr > F
Model	4	17785	4446.15952	116.68	<.0001
Error	47	1790.93411	38.10498		
Corrected Total	51	19576			

Root MSE	6.17292	R-Square	0.9085	
Dependent Mean	112.96923	Adj R-Sq	0.9007	
Coeff Var	5.46425			

Parameter Estimates

Variable	Label	DF	Parameter Estimate	Standard Error	t Value	Pr > |t|
Intercept	Intercept	1	-10.08918	7.35341	-1.37	0.1766
x1	TITLES	1	0.44914	0.25412	1.77	0.0836
x2	FOOTAGE	1	0.30836	0.18508	1.67	0.1024
x3	IBMBASE	1	0.06956	0.05110	1.36	0.1799
x4	APLBASE	1	-0.01260	0.08495	-0.15	0.8827

The partial slopes are all positive with the exception of the partial slope for APLBASE. The negative sign on the partial slope for APLBASE should not be of great concern because the standard error for the coefficient is much large in magnitude than the estimate of the coefficient. Therefore, the independent variable, APLBASE, provides very little additional predictive value given the other three independent variables in the model.

c. With $t_{0.025,47} = 2.012$ and $SE(\hat{\beta}_{TITLES}) = 0.25412$, a 95% C.I. for the coefficient associated with TITLES is given by

$$0.44914 \pm (2.012)(0.25412) \Rightarrow (-0.062, 0.960).$$

12.59   a. To test $H_o : \beta_1 = \beta_2 = \beta_3 = \beta_4 = 0$ versus $H_o :$ at least one $\beta_i \neq 0$, we can use the F-test from the SAS output: $F = 116.68$ with $p - value < 0.0001$. Yes, we can reject

123

H_o and conclude that at least one of the independent variables has predictive value for SALES.

b. The p-values for test $H_o : \beta_i = 0$ versus $H_a : \beta_i \neq 0$ for $i = 1, 2, 3, 4$ are 0.0836, 0.1024, 0.1799, and 0.8827. All of these values are relatively large (greater than $\alpha = 0.05$). Therefore, none of the independent variables add significant predictive value to a model that already contains the other three independent variables.

12.60 SAS output is given here:

The SAS System

The CORR Procedure

Pearson Correlation Coefficients, N = 52

	SALES	TITLES	FOOTAGE	IBMBASE	APLBASE
SALES	1.00000	0.94905	0.92648	0.92941	0.92476
TITLES	0.94905	1.00000	0.95623	0.96593	0.97277
FOOTAGE	0.92648	0.95623	1.00000	0.91200	0.93608
IBMBASE	0.92941	0.96593	0.91200	1.00000	0.94629
APLBASE	0.92476	0.97277	0.93608	0.94629	1.00000

All of the correlations between pairs of independent variables are very large (greater than 0.9). Thus the collinearity problem is very severe for this regression modelling. All four variables are related to the size of the store. For a store that is growing rapidly, all four of the independent variables would tend to increase in a similar fashion. Note also since this is time series data on a given store, the 52 values of the five variables would likely to be autocorrelated and hence violate the independence condition that is required in using the regression techniques for testing and model fitting.

12.61 From the model we have $R^2 = 0.9085$. The correlation between SALES and TITLES is 0.94905. It has square $(0.94905)^2 = 0.9007$. This corresponds to the R^2 for the one variable model relating SALES to just TITLES. Thus, the one variable model has nearly the same R^2 as the four variable model. The SAS output for the one variable model is given here:

The SAS System

Dependent Variable: y SALES

Analysis of Variance

Source	DF	Sum of Squares	Mean Square	F Value	Pr > F
Model	1	17632	17632	453.55	<.0001
Error	50	1943.77196	38.87544		
Corrected Total	51	19576			

```
Root MSE              6.23502    R-Square    0.9007
Dependent Mean      112.96923    Adj R-Sq    0.8987
Coeff Var             5.51922
```

Parameter Estimates

Variable	Label	DF	Parameter Estimate	Standard Error	t Value	Pr > |t|
Intercept	Intercept	1	-10.88845	5.87976	-1.85	0.0700
x1	TITLES	1	0.84633	0.03974	21.30	<.0001

In the complete model, we want to test

$H_o : \beta_2 = \beta_3 = \beta_4 = 0$ versus $H_a :$ at least one of $\beta_2, \beta_3, \beta_4 \neq 0$.

The F-statistic has the form:

$F = \frac{[17785 - 17632]/(4-1)}{1790.93411/[52-5]} = 1.34$

with $df = 3, 47 \Rightarrow p - value = Pr(F_{3,47} \geq 1.34) = 0.2727 \Rightarrow$ Fail to reject H_o.

There is not substantial evidence to conclude that at least one of $\beta_2, \beta_3, \beta_4 \neq 0$. Based on the F-test, omitting FOOTAGE, IBMBASE, and APLBASE from the model would not substantially changed the fit of the model. Dropping FOOTAGE, IBMBASE, and APLBASE from the model will not result in a large decrease in the predictive value of the model. These variables are so highly correlated with TITLES that they essentially add no additional predictive value once SALES is modelled by TITLES. Also, there is the possibility that the autocorrelation is confusing the issue. To truely relate SALES to the four independent variables we would need data from 52 different computer stores during at fixed time period.

12.63   a. The box plots of the data are given here:

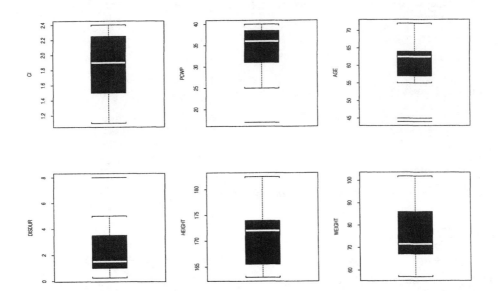

b. The scatterplots of the data are given here:

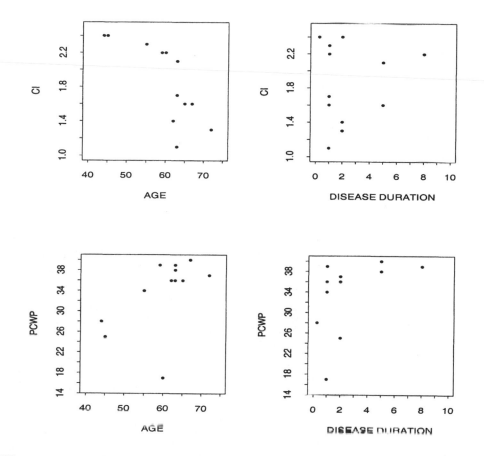

There appears to be somewhat of a negative correlation between Age and CI but a positive correlation between Age and PCWP. The relation between disease duration and CI is very weak but slightly positive. While the correlation between Disease Duration and PCWP is somewhat stronger.

12.64 For both CI and PCWP, the addition of the interaction term between Age and Disease Duration ($x_1 x_2$) did not appreciatively improve the fit of the model. The t-statistics for the interaction terms, which measures the effect of adding the interaction term to a model already containing the terms x_1 and x_2, were not significant. The p-values were 0.8864 for the CI model and 0.6838 for the PCWP model. The R^2 values for the two CI models were 67.39% for the model without the interaction term and 67.48% for the model with the interaction. Similarly, the R^2 values for the two PCWP models were 42% for the model without the interaction term and 43.27% for the model with the interaction. Thus, for both the CI model and the PCWP model the interaction term adds very little predictive value to the model. The CI model has a p-value of 0.0021 for the Age term but only 0.1514 for Disease Duration. This confirms the correlation depictions in the two scatterplots for CI, where Age appears to have a greater correlation with CI than the correlation between Disease Duration and CI. A similar relationship is observed in the PCWP model, although in this model neither Age nor Disease Duration appears to provide much predictive value for PCWP.

Chapter 13: More on Multiple Regression

13.2 Selecting the Variables (Step 1)

13.3 A matrixplot of the data is given here:

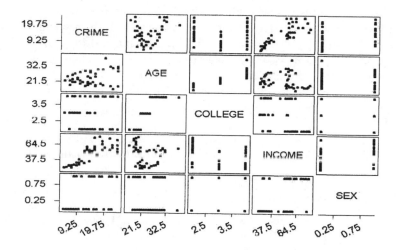

Based on the the scatterplots of the four independent variables, there does not appear to be a major problem with collinearity. The variables that would be included in the model are Age, Income, and Sex. This set of independent variables had a C_p value of 3.007 and $R^2 = 82.78394\%$, whereas the model with College, Age, Income, and Sex had a C_p value of 5.0 and $R^2 = 82.783774\%$. Thus, for the model with Age, Income, and Sex, the C_p is nearly equal to $p = 3$ and the change in R^2 by including College is nearly 0.

13.3 Formulating the Model (Step 2)

13.7 The fitted model is

$$\hat{y} = 288.062 + 174.815 \, \text{Log(DOSE)} - 0.344 \, \text{WEIGHT} - 0.863 \, \text{Log(DOSE)*WEIGHT}.$$

From the residual plots, there appears to be some lack of fit for the fitted model. However, note that the residuals are smaller than the residuals obtained from the model fit in Exercise 13.6. The value of R^2 has been increased from 30.24% to 55.85% and Root MSE reduced from 54.999 to 45.102.

13.9   a. Defining the "Industry" variable in this fashion would indicate that this variable is quantitative not qualitative. A one-unit increase in Industry could indicate a change from the chemical industry to the data-processing industry, or a change from the data-processing industry to the electronics industry. There is no logical reason to assume that these two possible changes would indicate the same change in the response variable, y. Thus, the coefficient associated with this variable would be meaningless.

   b. An improved approach would be to define three indicator variables:

$$x_1 = \begin{cases} 1 & \text{if} \quad \text{industry} = \text{chemical} \\ 0 & \text{if} \quad \text{Otherwise} \end{cases} \qquad x_2 = \begin{cases} 1 & \text{if} \quad \text{industry} = \text{data processing} \\ 0 & \text{if} \quad \text{Otherwise} \end{cases}$$

$$x_3 = \begin{cases} 1 & \text{if} \quad \text{industry} = \text{electronics} \\ 0 & \text{if} \quad \text{Otherwise} \end{cases}$$

   c. Another indicator variable could be defined to denote whether or not the firm matches employee contributions:

$$x_4 = \begin{cases} 1 & \text{if} \quad \text{firm matches employee contribution} \\ 0 & \text{if} \quad \text{Otherwise} \end{cases}$$

13.10   a. A model with CONTRIB as the dependent variable would generally just yield the conclusion that large companies contribute more money than small companies. This would not be too informative. Also, there may some collinearity problems due to the possible high correlation between INCOME and SIZE. However, this may not a problem if a wide variety of industries are included in the data base.

   b. The variable CONTRIB/INCOME would represent the proportion of pre-tax income contributed by the firm. This may be a more informative dependent variable.

13.11 This relationship could be modelled by including cross-product terms, such as

$$x_1 * \text{SIZE}, \quad x_2 * \text{SIZE}, \quad x_3 * \text{SIZE},$$

where x_1, x_2, x_3 are the indicator variables defined in Exercise 13.9, to differentiate the particular industries. These terms would tend yield unique partial slopes for SIZE for each of the four categories of INDUSTRY.

13.12   a. Such a relation between CONTRIB/INCOME and EDLEVEL is an indication of a nonlinear relation. Thus higher order terms in EDLEVEL such as $(\text{EDLEVEL})^2$ should be included in the model and then tested for significance.

     b. If the relation between CONTRIB/INCOME and EDLEVEL is as described then the residual plot from a first-order model in EDLEVEL would have a curved pattern (parabolic shape).

13.16   a. The correlation coefficients between the pairs of independent variables are all equal to 0. Thus, there is no simple collinearities in the data.

     b. Scatterplots of the pairs of independent variable are given here:

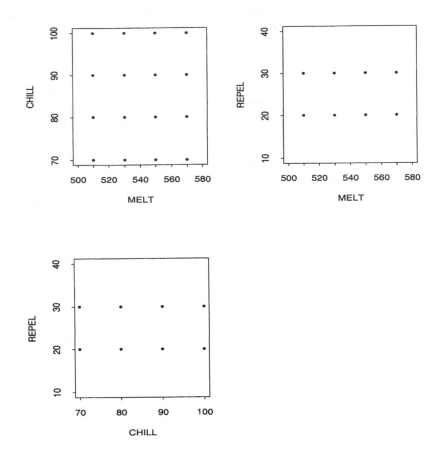

From the three scatterplots and because there are an equal number of data values at each plotted point, we observe that the correlation between MELT and REPEL, the correlation between MELT and CHILL, and the correlation between CHILL and REPEL are all 0 because the data points are equally placed on a grid. Thus, for example, an increase in CHILL from 80 to 90 at any value for MELT, is associated with both increases and decreases in MELT. A similar result holds for the other pairs. There is no possibility to predict MELT from given values of either CHILL or REPEL with the given data.

13.17 In each of the plots, there is no obvious pattern of nonlinearity. The residuals appear to be a random scatter about a horizontal line at height 0.

13.18   a. The R^2 has increased from 90.4% to 91.4%. Thus, there was very little change in the R^2 value.

     b. In the second-order model, we want to test
$H_o : \beta_6 = \beta_7 = 0$ versus H_a : at least one of $\beta_6, \beta_7 \neq 0$.

The F-statistic has the form:

$$F = \frac{[SS_{Reg.,Complete} - SS_{Reg.,Reduced}]/(k-g)}{SS_{Residual,Complete}/[n-(k+1)]} = \frac{[3141.625-3106.400]/(7-5)}{296.250/24} = 1.43$$

with $df = 2, 24 \Rightarrow p-value = Pr(F_{2,24} \geq 1.43) = 0.2590 \Rightarrow$

Fail to reject H_o. There is not substantial evidence to conclude that $\beta_6 \neq 0$ and/or $\beta_7 \neq 0$. Based on the F-test, omitting $(MELT)^2$ and $(CHILL)^2$ from the model would not substantially changed the fit of the model. The model with $(MELT)^2$ and $(CHILL)^2$ is not a significant improvement in fit over the fit of the first-order model.

   c. The p-values for testing whether $(MELT)^2$ or $(CHILL)^2$ are significant as the last predictor in the model are 0.429 and 0.151, respectively. Thus, neither term is significant as the last predictor in the model.

13.19   a. The independent variable entered into the model in the following order:

        Step 1: REPEL

        Step 2: KNIFE

        Step 3: CHILL

        Step 4: MELT

   b. The ordering (largest to smallest) of the independent variables based on the magnitude of their correlation with STIFF

        1. REPEL

        2. KNIFE

        3. CHILL

        4. MELT

        5. SPEED

   c. The two sets of orderings is the same. This would not happen in general but in this situation the independent variables have 0 pairwise correlation with each other. Thus, the next independent variable to enter the model will be the independent variable having highest correlation with the dependent variable, STIFF, irregardless of which of the other independent variables have already been entered in the model.

13.21   a. Test $H_o : \beta_1 = \beta_2 = \cdots = \beta_7 = 0$ versus H_a : at least one of $\beta_i's$ is not 0.

Test Statistic: $F = \dfrac{MS(Regression)}{MS(Error)} = 1.47$ with p-value=0.208.

Fail to reject H_o, there is not significant evidence that the seven independent variables provide predictive value for the dependent variable, CONTRIB/INCOME.

   b. The p-values for the the t-tests of the seven individual hypotheses: $H_a : \beta_i \neq 0$ are given on the output. Only SIZE with a p-value=0.028 has been shown to have statistical significance as the last variable to enter the model.

13.23   a. Nonconstant variance may be present; Weighted least squares or Transformation of the dependent variable is needed

      d. Potential serial correlation; Need to include terms in the model to account for the serial correlation.

13.24 The Minitab output of regressing y-differences on x-differences is given here:

```
The orginal data and first differences:

Row     y     x     yd    xd
 1     11   0.5     *     *
 2     10   1.0    -1    0.5
 3      2   1.2    -8    0.2
 4     14   1.4    12    0.2
 5     22   1.7     8    0.3
 6     10   1.8   -12    0.1
 7     20   2.0    10    0.2
 8     19   2.3    -1    0.3
 9     32   2.5    13    0.2
10     23   2.8    -9    0.3
11     40   3.0    17    0.2
12     37   3.1    -3    0.1
13     30   3.5    -7    0.4
14     43   3.6    13    0.1
15     55   3.8    12    0.2
16     29   4.2   -26    0.4
17     45   4.4    16    0.2
18     60   5.1    15    0.7
19     53   5.2    -7    0.1
20     30   5.4   -23    0.2
21     42   5.5    12    0.1
22     25   6.0   -17    0.5
23     63   6.2    38    0.2
24     51   6.3   -12    0.1
```

```
Regression Analysis: yd versus xd


The regression equation is
yd = 4.50 - 11.0 xd

23 cases used 1 cases contain missing values

Predictor      Coef     SE Coef       T       P
Constant      4.502       6.238     0.72   0.478
xd           -10.95       21.15    -0.52   0.610

S = 15.51      R-Sq = 1.3%      R-Sq(adj) = 0.0%

Analysis of Variance
```

Source	DF	SS	MS	F	P
Regression	1	64.5	64.5	0.27	0.610
Residual Error	21	5049.9	240.5		
Total	22	5114.4			

Unusual Observations

Obs	xd	yd	Fit	SE Fit	Residual	St Resid
18	0.700	15.00	-3.17	10.01	18.17	1.53 X
23	0.200	38.00	2.31	3.42	35.69	2.36R

R denotes an observation with a large standardized residual
X denotes an observation whose X value gives it large influence.

Durbin-Watson statistic = 2.88

Residuals Versus the Order of the Data

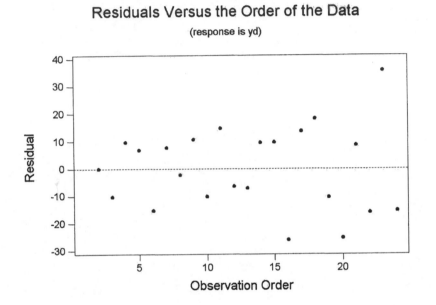

In examining the residual plot versus observation order, it would appear that a positive residual is followed by a negative residual is followed by a positive residual. Hence, there is an indication of negative serial correlation in the differenced data.

13.25 From the Minitab output Durbin-Watson statistic $= 2.88$. Because this value is greater than 2.5, there is an indication of negative correlation between y-differenced and x-differenced thus confirming what was observed in the residual plot.

13.27 When RATE5=9 and UNEMPLOY=16,

$$\hat{y} = -2.704 + (0.517)(9) + (1.450)(16) + (0.0353)(9)(16) = 30.23$$

The unemployment rate of 16% is considerably beyond the range of the data from which the model was fitted. Thus, an extrapolation problem may exist.

Supplementary Exercises

13.29 A scatterplot of the data is given here:

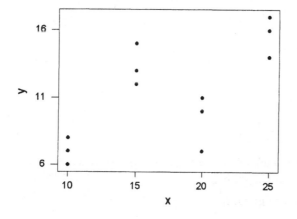

A model which includes linear, quadratic and cubic terms in x would appear to be needed:
$$y = \beta_o + \beta_1 x + \beta_2 x^2 + \beta_3 x^3 + \epsilon$$

13.30 Minitab output is given here:

```
Regression Analysis: y versus x, x^2, x^3


The regression equation is
y = - 119 + 24.3 x - 1.45 x^2 + 0.0276 x^3

Predictor        Coef     SE Coef        T       P        VIF
Constant      -119.33       24.85    -4.80   0.001
x              24.344       4.766     5.11   0.001     3406.9
x^2           -1.4467      0.2864    -5.05   0.001    15313.5
x^3          0.027556    0.005443     5.06   0.001     4427.8

S = 1.581           R-Sq = 87.2%        R-Sq(adj) = 82.4%
PRESS = 45.0000     R-Sq(pred) = 71.28%

Analysis of Variance

Source            DF          SS          MS       F       P
Regression         3     136.667      45.556   18.22   0.001
Residual Error     8      20.000       2.500
Total             11     156.667

The number of distinct predictor combinations equals the
number of parameters. No degrees of freedom for lack of fit.

Cannot do pure error test.

Source      DF      Seq SS
x            1      72.600
x^2          1       0.000
x^3          1      64.067


Durbin-Watson statistic = 2.41
```

a. The fitted regression equation is
$$\hat{y} = -119 + 24.3x - 1.45x^2 + 0.0276x^3$$

b. It is not possible to test for lack of fit since the number of distinct values of x equals the number of parameters in the model.

c. Residual plots of the data are given here:

Residuals Versus the Fitted Values
(response is y)

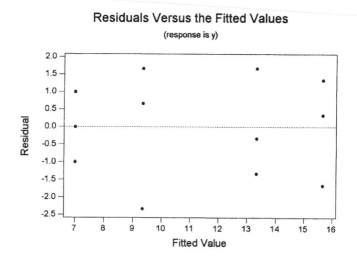

An examination of the residual plot does not reveal any obvious violations of the model conditions.

13.31 Minitab output is given here:

```
(EX13-31)        Regression Analysis: y versus x, x^2, x^3


The regression equation is
y = - 91.0 + 19.2 x - 1.16 x^2 + 0.0227 x^3

Predictor        Coef      SE Coef         T        P        VIF
Constant       -91.00        35.89     -2.54    0.064
x              19.217         6.820      2.82    0.048     4751.4
x^2           -1.1633        0.4016     -2.90    0.044    19931.4
x^3          0.022667      0.007473      3.03    0.039     5413.7
```

139

```
S = 1.555              R-Sq = 90.7%          R-Sq(adj) = 83.7%
PRESS = *              R-Sq(pred) =    *%
```

Analysis of Variance

```
Source             DF        SS         MS        F        P
Regression          3      94.333     31.444    13.01    0.016
Residual Error      4       9.667      2.417
Total               7     104.000
```

The number of distinct predictor combinations equals the
number of parameters. No degrees of freedom for lack of fit.

Cannot do pure error test.

```
Source        DF      Seq SS
x              1      58.329
x^2            1      13.773
x^3            1      22.231
```

Unusual Observations
```
Obs       x          y        Fit      SE Fit    Residual    St Resid
 3      15.0     12.000     12.000     1.555       0.000         * X
```

X denotes an observation whose X value gives it large influence.

Durbin-Watson statistic = 2.32

a. The fitted regression equation is
$$y = -91.0 + 19.2x - 1.16x^2 + 0.0227x^3$$

b. Residual plots of the data are given here:

Residuals Versus the Fitted Values
(response is y)

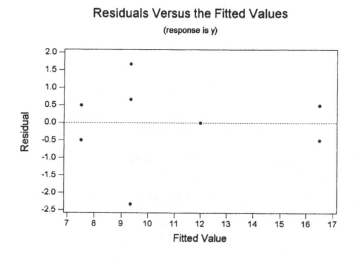

The two model have equivalent fits to the data. The model fit to the data set having the four data values deleted appears to have a single outlier at x=20, y=7.

13.33   a. See output on page 787.

b. $\hat{y} = 1.20 + 7.021$ LOG(DOSE)

c. See output.

d. The model using LOG(DOSE) appears to provide the better fit:
Larger R^2: 92.95% vs 88.15% for Quadratic, 77.30% for Linear
Smaller MS(Error): 4.144 vs 7.548 for Quadratic, 13.345 for Linear
Residual Plots: No apparent pattern for LOG(DOSE) model, whereas both the Quadratic and Linear models have somewhat of a parabolic pattern.

13.37 The Minitab output is given here:

```
The regression equation is
y = 60.5 - 0.705 x1 + 0.00328 x1^2 + 8.88 x2

Predictor        Coef      SE Coef         T        P       VIF
Constant       60.477        5.512     10.97    0.000
x1            -0.70507      0.07551     -9.34    0.000     106.5
x1^2         0.0032768    0.0002505     13.08    0.000     106.5
x2              8.875        2.499       3.55    0.001       1.0

S = 1.732              R-Sq = 97.4%       R-Sq(adj) = 97.2%
PRESS = 165.551        R-Sq(pred) = 96.70%

Analysis of Variance

Source            DF        SS         MS         F        P
Regression         3     4877.7     1625.9     542.22    0.000
Residual Error    44      131.9        3.0
  Lack of Fit      8       44.2        5.5       2.27     0.044
  Pure Error      36       87.7        2.4
Total             47     5009.6
-----------------------------------------------------------

Regression Analysis: y versus x1, x1^2, x2, x1*x2, x1^2*x2

The regression equation is
y = 42.3 - 0.421 x1 + 0.00224 x1^2 + 69.5 x2 - 0.949 x1*x2 + 0.00345 x1^2*x2

Predictor        Coef      SE Coef         T        P       VIF
Constant        42.28        17.01       2.49    0.017
x1             -0.4205       0.2351      -1.79    0.081    1064.7
x1^2         0.0022422    0.0007800      2.87    0.006    1064.7
x2              69.54        53.77       1.29    0.203     477.4
x1*x2          -0.9485       0.7435      -1.28    0.209    3118.0
x1^2*x2       0.003449     0.002467      1.40    0.169    1627.3

S = 1.705              R-Sq = 97.6%       R-Sq(adj) = 97.3%
PRESS = 178.669        R-Sq(pred) = 96.43%

Analysis of Variance

Source            DF        SS         MS         F        P
Regression         5     4887.53     977.51    336.22    0.000
Residual Error    42      122.11       2.91
  Lack of Fit      6       34.42       5.74      2.35     0.051
  Pure Error      36       87.69       2.44
Total             47     5009.64
```

The first fitted regression equation is

$$\hat{y} = 60.477 - 0.705x1 + 0.00328x1^2 + 8.875x2$$

The second fitted regression equation is

$$\hat{y} = 42.28 - 0.421x1 + 0.00224x1^2 + 69.54x2 - 0.949x1x2 + 0.00345x1^2x2$$

These two models provide only marginal improvement over the quadratic model in just x_1. However, the pattern in the residual plot noted from the quadratic model in x_1 is not as noticeable in the residual plots from these two models.

A residual plot of the first fitted model is given here:

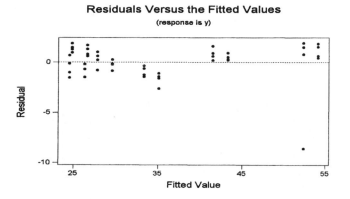

13.38　a. A scatterplot of the data is given here:

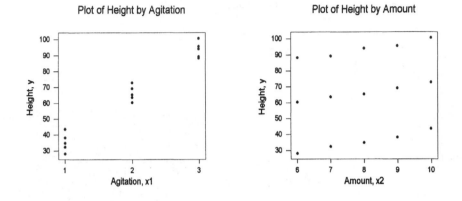

Plot of Height by Agitation

Plot of Height by Amount

The model suggested by the scatterplot is
$$y = \beta_o + \beta_1 x_1 + \beta_2 x_1^2 + \beta_3 x_1 x_2 + \beta_4 x_1^2 x_2$$

b. There is no indication of normality in the data.

c. Minitab output is given here:

```
The regression equation is
Height, y = - 29.9 + 38.6 x1 + 3.84 x2 - 1.82 x1^2
            - 0.275 x1*x2

Predictor       Coef     SE Coef        T        P       VIF
Constant     -29.873       3.866    -7.73    0.000
Agitatio      38.580       2.542    15.18    0.000      81.0
Amount,       3.8367       0.4315     8.89    0.000       7.0
x1^2         -1.8200       0.4892    -3.72    0.004      49.0
x1*x2        -0.2750       0.1997    -1.38    0.199      39.0

S = 0.8932          R-Sq = 99.9%       R-Sq(adj) = 99.9%
PRESS = 21.3500     R-Sq(pred) = 99.76%
```

```
Analysis of Variance

Source          DF        SS        MS        F        P
Regression      4       8804.7    2201.2   2759.00   0.000
Residual Error  10         8.0       0.8
Total           14      8812.7
```

The fitted model is given here:

$$\hat{y} = -29.873 + 38.580x1 + 3.837x2 - 1.820x1^2 - 0.275x1x2$$

d. A residual plot of the first fitted model is given here:

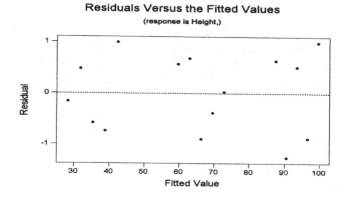

Residuals Versus the Fitted Values
(response is Height,)

There is no apparent pattern in the data. The condition of constant variance does not appear to be violated.

13.39 The is no indication of the plot of Height by Amount of a quadratic curvature. Hence, the second order terms in Amount is probably unnecessary.

13.41   a. The model is given on page 735:

$$y = \beta_0 + \beta_1 x_1 + \beta_2 x_1^2 + \beta_3 x_2 + \beta_4 x_2^2 + \beta_5 x_3 + \beta_6 x_1 x_3 + \beta_7 x_1^2 x_3 + \beta_8 x_2 x_3 + \beta_9 x_2^2 x_3 + \epsilon$$

β_o is the y-intercept for the Standard Model

β_1 is partial slope of Price per Gallon for Standard Model

β_2 is partial slope of (Price per Gallon)^2 for Standard Model

β_3 is partial slope of Interest Rate for Standard Model

β_4 is partial slope of (Interest Rate)^2 for Standard Model

β_5 is difference in the y-intercept between Standard and Luxury Models

β_6 is difference partial slopes of Price per Gallon between Standard and Luxury Models

β_7 is difference partial slopes of (Price per Gallon)^2 between Standard and Luxury Models

β_8 is difference partial slopes of Interest Rate between Standard and Luxury Models

β_9 is difference partial slopes of (Interest Rate)^2 between Standard and Luxury Models

b. The estimated model is given by

$\hat{y} = -3.218 + 28.495x_1 - 7.966x_1^2 - 3.490x_2 + 0.154x_2^2 - 300.022x_3 + 389.508x_1x_3 - 112.642x_1^2x_3 - 5.030x_2x_3 + 0.217x_2^2x_3$

For Luxury: $x_3 = 0 \Rightarrow \hat{y} = -3.218 + 28.495x_1 - 7.966x_1^2 - 3.490x_2 + 0.154x_2^2$

For Standard $x_3 = 1 \Rightarrow \hat{y} = -303.240 + 418.003x_1 - 120.608x_1^2 - 8.520x_2 + 0.371x_2^2$

c. The regression coefficients for the two models are considerably different. However, the standard errors for the interaction terms involving the dummy variable for Type of Car are very large. This may result in a finding that the two models are not significantly different. A reduced model in which the interaction terms are dropped should be run in order to determine if the two models are in fact significantly different.

13.43 Minitab output is given here:

```
The regression equation is
y = 44.2 - 0.494 x + 0.00143 x^2

Predictor        Coef      SE Coef        T         P
Constant       44.182        1.756     25.16     0.000
x             -0.49400      0.06688     -7.39     0.000
x^2           0.0014285    0.0005163      2.77     0.017

S = 3.273       R-Sq = 96.0%      R-Sq(adj) = 95.3%

Analysis of Variance

Source            DF          SS          MS         F         P
Regression         2        3071.2      1535.6     143.33     0.000
Residual Error    12         128.6        10.7
  Lack of Fit      3          36.7        12.2       1.20     0.364
  Pure Error       9          91.8        10.2
Total             14        3199.7
```

a. The fitted model is $\hat{y} = 44.182 - 0.494x + 0.00143x^2$

b. From the output $F = \frac{36.7/3}{91.8/9} = 1.20 \Rightarrow p-value = 0.364$

Thus, there is not significant evidence of lack of fit of the model. Thus higher order terms in Temperature (x) are not needed to adequately fit the data.

c. A residual plot of the fitted data is given here:

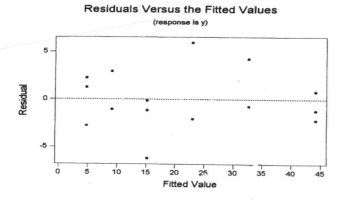

Residuals Versus the Fitted Values
(response is y)

There is no obvious patterns in the residual plot.

13.47   a. A scatterplot of the data is given in the textbook.

b. The estimated linear regression equation is given here:
$$\hat{y} = -1.540 + 0.70635x$$

c. The estimated quadratic regression equation is given here:
$$\hat{y} = 9.179 - 0.0468x + 0.011587x^2$$

d. Using Linear equation, when Temperature is $27°C$, $\hat{y} = 17.5$.
Using Quadratic equation, when Temperature is $27°C$, $\hat{y} = 16.4$.

13.53 Test the hypotheses, $H_o : \beta_{AGE} \geq -2500$ vs $H_a : \beta_{AGE} < -2500$

$$t = \frac{\hat{\beta}_{AGE} - 2500}{SE(\hat{\beta}_{AGE})} = \frac{-506 - (-2500)}{1111} = 1.79, \text{ with df=12.}$$

$$p - value = Pr(t_{12} \leq 1.79) = 0.9507 \Rightarrow$$

Fail to reject H_o. There is not evidence that the depreciation per year is less than $2500.

13.55 The following table summaries the fit of the many models. The p-values for comparing the reduced model to the full model are also given, along with the maximum p-value for testing whether to drop any individual term in the given model.

Number of Ind. Var. in Model	R^2_{adj}	MS(Residual)	$df_{Res.}$	p-value for Reducing Model	Var. Removed from Model
9	47.2%	273282828	12	0.898	BEDB, AGE
7	53.9%	238486598	14	0.412	BEDA
6	54.8%	233974343	15	0.346	BATHS
5	54.9%	233174830	16	0.136	CARB
4	51.1%	253230451	17	0.188	LOT
3	48.7%	265619875	18	0.081	DOM
2	42.1%	299566628	19	0.013	CARA
1	23.0%	398320115	20	0.014	BEDC

The "best" model is the model with the independent variables: DOM, CARA, and BEDC. It has a considerably larger value for R^2_{adj} than the two-variable model without DOM, the p-value for dropping DOM from this model is only 0.081, which is marginally significant, it has a $C_p = 3.5$ which is fairly close to $p = 3$ and there is a considerable increase in MS(Residual) when DOM is dropped from the model.

13.57   a. The question is a test of $H_o : \beta_1 = \beta_2 = 0$ vs $H_a : \beta_1 \neq 0$ and/or $\beta_2 \neq 0$.

From the output, $F = \frac{MS(Model)}{MS(Error)} = 15.987$, with $p - value < 0.0001 < 0.05 \Rightarrow$

Reject H_o and conclude there is significant evidence that ROOMS and SQFT taken together contain information about PRICE.

   b. Test $H_o : \beta_1 = 0$ vs $H_a : \beta_1 \neq 0$.

$t = 0.717$ with $p - value = 0.4822 > 0.05 \Rightarrow$

Fail to reject H_o and conclude there is not significant evidence that the coefficient of ROOMS is different from 0.

   c. Test $H_o : \beta_2 = 0$ vs $H_a : \beta_2 \neq 0$.

$t = 0.1.468$ with $p - value = 0.1585 > 0.05 \Rightarrow$

Fail to reject H_o and conclude there is not significant evidence that the coefficient of SQFT is different from 0.

13.59 The $F - test$ for the overall model is 4.42 with p-value=0.0041.

The indicator variable DV_3 measures the difference in risk of infection between hospital in the South and West holding all other variables constant. The coefficient of DV_3 is β_7 and we want to test $H_o : \beta_7 \leq .5\%$ vs $H_a : \beta_7 > .5\%$. The test statistic is

$t = \frac{\hat{\beta}_7 - 0.5}{SE(\hat{\beta}_7)} = \frac{0.7024 - 0.5}{0.8896} = 0.23$ with df=20

$p - value = Pr(t_{20} > 0.23) = 0.4102 \Rightarrow$

Fail to reject H_o, there is not significant evidence that the infection rate in the south is at least 0.5% higher than in the west.

13.61 The table on page 818 provides summary information for a one variable at a time elimination from the full model. The following model is selected based on this information:

$y = \beta_0 + \beta_1 \text{ STAY} + \beta_3 \text{ RCR} + \epsilon$

The R^2 for this model is 0.5578 vs 0.6072 for the seven variable model.

The MS(Error) for this model is 28.765 vs 25.546 for the seven variable model.

A test of H_o : Two Variable Model vs H_o : Seven Variable Model is given by testing the following parameters in the seven variable model:

$H_o : \beta_2 = \beta_4 = \beta_5 = \beta_6 = \beta_7 = 0$ vs H_a : at least one of $\beta_2, \beta_4, \beta_5, \beta_6, \beta_7 \neq 0$

$F = \frac{(39.49805177 - 36.27961297)/5}{25.54623394/20} = 0.50$ with $df = 5, 20 \Rightarrow$

$p - value = Pr(F_{5,20} > 0.50) = 0.7726 \Rightarrow$

Fail to reject H_o, there is not significant evidence that any of the five parameters are not 0. Thus, there is not significant evidence of a difference between the two-variables and seven-variables models.

Based on the above test, the marginal difference in R^2 and MS(Error), the model with fewer variables is the more desirable model.

13.63   a. The model F-test has value 6.52 with p-value=0.0004. Thus, we can reject $H_o : \beta_1 = \beta_2 = \beta_3 = \beta_4 = \beta_5 = \beta_6 = 0$ and conclude that at least one of $\beta_1, \beta_2, \beta_3, \beta_4, \beta_5, \beta_6 \neq 0$. There is significant evidence that the set of six independent variables does provide an explaination of the variability in PULSE.

   b. The pairwise correlations between the six independent variables ranged from -0.74863 to 0.59885. Thus, there may be a degree of multicollinearity due to pairwise correlations. The largest correlation was between PHYS1 and PHYS2. Multicollinearity may greatly inflate the standard errors of the predictions.

   c. The coefficient of the variable PHYS1 measures the difference in average pulse between individuals who have substantial physical exercise and those who exercise physical. The estimate of this parameter is 13.43 which indicates individuals who exercise substantially have a mean increase in pulse rate of 13.43 units higher than those who exercise very little.

   A 95% C.I. for the coefficient of PHYS1 is

   $13.43 \pm (t_{.025,23})(SE(\beta)) \Rightarrow 13.43 \pm (2.069)(4.2512) \Rightarrow (4.63, 22.23)$

   Thus, we are 95% confident that the actual average difference in the increase in pulse between the two groups is greater than zero.

13.65   a. Based on the information contained in the model fit summary table, dropping HEIGHT and WEIGHT from the original six variable did not substantially affect the fit of the model.

   Six-variable model has $C_p = 7 > 6 = p$, whereas, the Four-variable model has $C_p = 3.26 \approx 4 = p$,

   Six-variable model has MS(Error)=47.3, which is larger than the value for the Four-variable model: MS(Error)=44.02

   Six-variable model has $R^2 = 0.6297$, which is only slightly larger than the value for the Four-variable model: $R^2 = 0.6255$.

   The best model amongest models having similar values for the criterions used to evaluate the fit of the model is the model having fewest independent variables. Thus, the four-variable model, excluding HEIGHT and WEIGHT is the best model.

   b. The model with interactions between all pairs of qualitative variables would be

   $y = \beta_0 + \beta_1 \text{ RUN} + \beta_2 \text{ SMOKE} + \beta_3 \text{ PHY1} + \beta_4 \text{ PHY2} + \beta_5 \text{ RUN*SMOKE} + \beta_6 \text{ RUN*PHY1} + \beta_7 \text{ RUN*PHY2} + \beta_8 \text{ SMOKE*PHY1} + \beta_9 \text{ SMOKE*PHY2} + \epsilon$

   There is no term for $PHY1 * PHY2$ since there is no meaning to such a term.

13.66   a. A scatterplot of the data is given here:

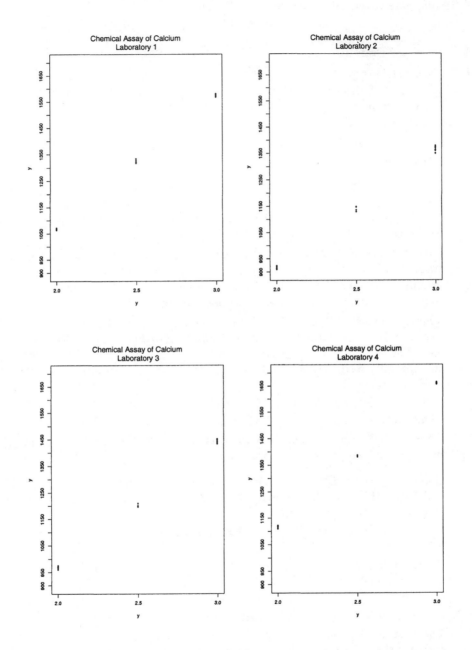

b. The following Minitab output contains the equations and predictions with the standard deviations of the four predictions.

```
Regression Analysis: y versus x for Laboratory 1


The regression equation is
y = 53.925 + 507.250 x
```

Row	Lab	x	y	y1	Sample	Pred x	StDevPred x
1	1	2.0	1068	1206	W1	2.27122	0.0044
2	1	2.0	1071	1202	W1	2.26333	*
3	1	2.0	1067	1202	W1	2.26333	*
4	1	2.0	1066	1201	W1	2.26136	*
5	1	2.0	1072	1194	W2	2.24756	0.0081
6	1	2.0	1068	1193	W2	2.24559	*
7	1	2.0	1064	1189	W2	2.23770	*
8	1	2.0	1067	1185	W2	2.22982	*
9	1	2.5	1333	1387	U1	2.62804	0.0065
10	1	2.5	1321	1387	U1	2.62804	*
11	1	2.5	1326	1384	U1	2.62213	*
12	1	2.5	1317	1380	U1	2.61424	*
13	1	3.0	1579	1394	U2	2.64184	0.0156
14	1	3.0	1576	1390	U2	2.63396	*
15	1	3.0	1578	1383	U2	2.62016	*
16	1	3.0	1572	1370	U2	2.60636	*
17	1	3.0	1579	1478	Y1	2.80744	0.0123
18	1	3.0	1571	1480	Y1	2.81138	*
19	1	3.0	1579	1473	Y1	2.79759	*
20	1	3.0	1567	1466	Y1	2.78379	*
21	1	*	*	1483	Y2	2.81730	0.0100
22	1	*	*	1477	Y2	2.80547	
23	1	*	*	1482	Y2	2.81533	
24	1	*	*	1472	Y2	2.79561	

Regression Analysis: y versus x for Laboratory 2

The regression equation is
y = 19.638 + 448.125 x

Row	Lab	x	y	y1	Sample	Pred x	StDevPred x
1	2	2.0	910	1017	W1	2.22563	0.0074
2	2	2.0	916	1017	W1	2.22563	*
3	2	2.0	915	1012	W1	2.21448	*
4	2	2.0	915	1020	W1	2.23233	*
5	2	2.0	913	1012	W2	2.21448	0.0105
6	2	2.0	923	1018	W2	2.22786	*
7	2	2.0	914	1015	W2	2.22117	*
8	2	2.0	921	1023	W2	2.23902	*
9	2	2.5	1129	1188	U1	2.60722	0.0135
10	2	2.5	1148	1199	U1	2.63177	*
11	2	2.5	1136	1197	U1	2.62731	*
12	2	2.5	1147	1202	U1	2.63846	*
13	2	3.0	1359	1186	U2	2.60276	0.0124
14	2	3.0	1378	1196	U2	2.62508	*
15	2	3.0	1370	1193	U2	2.61838	*
16	2	3.0	1373	1199	U2	2.63177	*
17	2	3.0	1349	1263	Y1	2.77459	0.0186
18	2	3.0	1361	1280	Y1	2.81252	*
19	2	3.0	1359	1280	Y1	2.81252	*
20	2	3.0	1363	1279	Y1	2.81029	*
21	2	*	*	1259	Y2	2.76566	0.0109
22	2	*	*	1269	Y2	2.78798	
23	2	*	*	1259	Y2	2.76566	
24	2	*	*	1265	Y2	2.77905	

Regression Analysis: y versus x for Laboratory 3

The regression equation is
y = 19.438 + 473.625 x

Row	Lab	x	y	y1	Sample	Pred x	StDevPred x
1	3	2.0	969	1090	W1	2.26036	0.0111
2	3	2.0	975	1098	W1	2.27725	*
3	3	2.0	969	1090	W1	2.26036	*
4	3	2.0	972	1100	W1	2.28147	*
5	3	2.0	969	1088	W2	2.25614	0.0062
6	3	2.0	960	1092	W2	2.26458	*
7	3	2.0	960	1087	W2	2.25402	*
8	3	2.0	966	1085	W2	2.24980	*
9	3	2.5	1196	1270	U1	2.64041	0.0104
10	3	2.5	1196	1261	U1	2.62140	*
11	3	2.5	1209	1261	U1	2.62140	*
12	3	2.5	1200	1269	U1	2.63829	*
13	3	3.0	1451	1261	U2	2.62140	0.0108
14	3	3.0	1440	1268	U2	2.63618	*
15	3	3.0	1439	1270	U2	2.64041	*
16	3	3.0	1449	1273	U2	2.64674	*
17	3	3.0	1439	1352	Y1	2.81354	0.0095
18	3	3.0	1433	1349	Y1	2.80720	*
19	3	3.0	1433	1353	Y1	2.81565	*
20	3	3.0	1445	1343	Y1	2.79454	*
21	3	*	*	1349	Y2	2.80720	0.0063
22	3	*	*	1353	Y2	2.81565	
23	3	*	*	1349	Y2	2.80720	
24	3	*	*	1355	Y2	2.81987	

Regression Analysis: y versus x for Laboratory 4

The regression equation is
y = 30.414 + 542.875 x

Row	Lab	x	y	y1	Sample	Pred x	StDevPred x
1	4	2.0	1122	1256	W1	2.25759	0.0073
2	4	2.0	1117	1254	W1	2.25390	*
3	4	2.0	1119	1256	W1	2.25759	*
4	4	2.0	1120	1263	W1	2.27048	*
5	4	2.0	1122	1260	W2	2.26495	0.0116
6	4	2.0	1110	1251	W2	2.24838	*
7	4	2.0	1111	1252	W2	2.25022	*
8	4	2.0	1116	1264	W2	2.27232	*
9	4	2.5	1386	1453	U1	2.62047	0.0063
10	4	2.5	1381	1447	U1	2.60942	*
11	4	2.5	1381	1451	U1	2.61678	*
12	4	2.5	1387	1455	U1	2.62415	*
13	4	3.0	1656	1450	U2	2.61494	0.0088
14	4	3.0	1663	1446	U2	2.60757	*
15	4	3.0	1659	1448	U2	2.61126	*
16	4	3.0	1665	1457	U2	2.62784	*
17	4	3.0	1658	1543	Y1	2.78625	0.0044
18	4	3.0	1658	1548	Y1	2.79546	*
19	4	3.0	1661	1543	Y1	2.78625	*
20	4	3.0	1660	1545	Y1	2.78994	*
21	4	*	*	1545	Y2	2.78994	0.0031
22	4	*	*	1546	Y2	2.79178	
23	4	*	*	1548	Y2	2.79546	
24	4	*	*	1544	Y2	2.78809	

c. Laboratory 4 has the smallest variation in its predictions, overall. However, for W, Lab 1 has the smallest variation, for U, Lab 4 has the smallest variation, and for Y, Lab 4 has the smallest variation.

13.67   a. No, the linear regression lines will not change because they are based on the mean of the y-values.

     b. Predictions of x will change because we are now predicting means of four measurements, rather than individual values.

Chapter 14: Design Concepts for Experiments and Studies

Exercises

14.1 a. Water Temperature and Type of Hardener

b. Water temperature: $175°F$ and $200°F$
   Type of Hardener: H_1, H_2, H_3

c. Manufacuring Plants

d. Plastic Pipe

e. Location on Plastic Pipe

f. 2 Pipes per Treatment

g. None

h. 6 Treatments: $(175°F, H_1), (175°F, H_2), (175°F, H_3)$
   $(200°F, H_1), (200°F, H_2), (200°F, H_3)$

14.3 a. • Factors: Location in Orchard, Location on Tree, Time of Year
   • Factor Levels: Location in Orchard: 8 Sections
      Time of Year: Oct., Nov., Dec., Jan., Feb., March, April, May
      Location on Tree: Top, Middle, Bottom
   • Blocks: None
   • Experimental Units: Locaton on Tree during one of the 8 months
   • Measurement Units: Oranges
   • Replications: For each section, time of year, location on tree, there is one experimental unit. Hence 1 rep.
   • Covariates: None
   • Treatments: 192 combinations of 8-Sections, 8 Months, 3 Locations on Tree yield (S_i, M_j, L_k), for $i = 1, \cdots, 8; j = 1, \ldots, 8; k = 1, \ldots, 3$

c. • Factors: Type of Treatment
   • Factor Levels: T_1, T_2
   • Blocks: Hospitals, Wards
   • Experimental Units: Patients
   • Measurement Units: Patients
   • Replications: 2 Patients per Treatment in each of the Ward/Hospital combinations
   • Covariates: None
   • Treatments: T_1, T_2

14.5 The necessary parameters are $t = 4, D = 30, \alpha = .05, \sigma = 12.25 \Rightarrow$

$$\phi = \sqrt{\frac{r(30)^2}{(2)(4)(12.25)^2}} = .8658\sqrt{r}.$$

Determine r so that power is .90. Select values for r compute $\nu_1 = t - 1 = 4 - 1 = 3, \nu_2 = t(r-1) = 4(r-1)$, and $\phi = .8658\sqrt{r}$, then use Table 14 with $\alpha = .05$ and $t = 4$ to determine power:

r	ν_2	ϕ	Power
5	16	1.94	.84
6	20	2.12	.91

Thus, it would take 6 reps to obtain a power of at least .90.

14.9   a. Bake one cake from each recipe in the oven at the same time. Repeat this procedure r times. The baking period is a block with the four treatments (recipes) appearing once in each block. The four recipes should be randomly assigned to the four positions, one cake per position. Repeat this procedure r times.

b. If position in the oven is important, then Position in the oven is a second blocking factor along with the baking Period. Thus, we have a latin square design. To have $r = 4$, we would need to have each recipe appear in each Position exactly once within each of four baking Periods. For example:

Period 1		Period 2		Period 3		Period 4	
R_1	R_2	R_4	R_1	R_3	R_4	R_2	R_3
R_3	R_4	R_2	R_3	R_1	R_2	R_4	R_1

c. We now have an incompleteness in the blocking variable Period since only four of the five recipes can be observed in each Period. In order to achieve some level of balance in the design, we need to select enough Periods in order that each recipe appears the same number of times in each period and the same total number of times in the complete experiment. For example, suppose we wanted to observe each recipe $r = 4$ times in the experiment. It would be necessary to have five periods in order to observe each recipe four times in each in of the four positions with exactly four recipes observed in each of the five periods.

Period 1		Period 2		Period 3		Period 4		Period 5	
R_1	R_2	R_5	R_1	R_4	R_5	R_3	R_4	R_2	R_3
R_3	R_4	R_2	R_3	R_1	R_2	R_5	R_1	R_4	R_5

14.11   a. Design B. The experimental units are not homogeneous since one group of consumers gives uniformly low scores and another group gives uniformly high scores, no matter what recipe is used. Using Design A, it is possible to have a group of consumers which give mostly low scores randomly assigned to a particular recipe. This would bias this particular recipe. Using Design B, the experimental error would be reduced since each consumer would evaluate each recipe. That is, each consumer is a block and each of the treatments (recipes) are observed in each block. This results in having each recipe subject to consumers who give low scores and to consumers who give high scores.

b. This would not be a problem for either design. In design A, each of the remaining four recipes would still be observed by 20 consumers. In design B, each consumer would still evaluate each of the four remaining recipes.

Chapter 15: Analysis of Variance for Standard Designs

15.3 Randomized Complete Block Design

15.1   a. The F-test from the ANOVA table tests 2-sided alternatives:

Test $H_o : \mu_{Attend} = \mu_{DidNot}$ vs $H_o : \mu_{Attend} \neq \mu_{DidNot}$

The ANOVA table is given here:

Source	DF	SS	MS	F	p-value
Pair	5	1319.42	263.88		
Treatment	1	420.08	420.08	71.40	0.0004
Error	5	29.42	5.88		
Total	11	1768.92			

Reject H_o and conclude there is significant evidence that the mean scores of students attending Head Start are significantly different from the mean scores of students who do not attend Head Start.

b. $RE(RCB,CR) = \frac{(b-1)MSB + b(t-1)MSE}{(bt-1)MSE} = \frac{(6-1)(263.88) + (6)(2-1)(5.88)}{((6)(2)-1)(5.88)} = 20.94 \Rightarrow$

It would take approximately 21 times as many observations (126) per treatment in a completely randomized design to achieve the same level of precision in estimating the treatment means as was accomplished in the randomized complete block design.

15.2   The F-test from the ANOVA table tests 2-sided alternatives

$H_o : \mu_{Attend} = \mu_{DidNot}$ vs $H_o : \mu_{Attend} \neq \mu_{DidNot}$.

The paired t-test can test one-sided alternatives:

$H_o : \mu_{Attend} \leq \mu_{DidNot}$ vs $H_o : \mu_{Attend} \geq \mu_{DidNot}$

$\bar{X}_{Attend} = 78.33 \quad \bar{X}_{DidNot} = 66.50$

$t = \frac{\bar{D}}{S_D / \sqrt{n}} = \frac{11.83}{3.43 / \sqrt{6}} = 8.45$

$p - value = Pr(t_5 \geq 8.45) = 0.0002 \Rightarrow$

Yes, there is significant evidence that the mean scores attending Head Start are greater than the mean scores of students who do not attend.

Note $(8.45)^2 = 71.40$

15.5   a. $y_{ij} = \mu + \alpha_i + \beta_j + \epsilon_{ij}; \quad i = 1, 2, 3, \quad j = 1, 2, 3, 4, 5, 6, 7$

y_{ij} is score on test of jth subject hearing the ith music type

α_i is the ith music type effect

β_j is the jth subject effect

$\hat{\mu} = 21.33, \quad \hat{\alpha}_1 = -0.47, \quad \hat{\alpha}_2 = -1.19, \quad \hat{\alpha}_3 = 1.67$

$\hat{\beta}_1 = 0, \quad \hat{\beta}_2 = -3, \quad \hat{\beta}_3 = 3.33, \quad \hat{\beta}_4 = -1.33, \quad \hat{\beta}_5 = 1, \quad \hat{\beta}_6 = 3.67, \quad \hat{\beta}_7 = -3.67$

b. $F = \frac{SS_{TRT}/df_{TRT}}{SS_{Error}/df_{Error}} = \frac{30.952/2}{28.38/12} = 6.54$ with df=2,12.

Therefore, $p - value = Pr(F_{2,12} \geq 6.54) = 0.0120 \Rightarrow$ Reject $H_o : \mu_1 = \mu_2 = \mu_3$.

We thus conclude that there is significant evidence of a difference in mean typing scores for the three types of music.

c. An interaction plot of the data is given here:

Interaction Plots

Based on the interaction plot, the additive model may be inappropriate because there is some crossing of the three lines. However, the plotted points are means of a single observation and hence may be quite variable in their estimation of the population means μ_{ij}. Thus, exact parallelism is not required in the data plots to ensure the validity of the additive model.

d. $t = 3, b = 7 \Rightarrow RE(RCB, CR) = \frac{(7-1)(24.889)+(7)(3-1)(2.365)}{((7)(3)-1)(2.365)} = 3.86 \Rightarrow$

It would take 3.86 times as many observations (approximately 27) per treatment in a completely randomized design to achieve the same level of precision in estimating the treatment means as was accomplished in the randomized complete block design. Since RE was much larger than 1, we would conclude that the blocking was effective.

15.4 Latin Square Design

15.9 a. $y_{ij} = \mu + \alpha_k + \beta_i + \gamma_j + \epsilon_{ij}$; $i, j, k = 1, 2, 3, 4$;

where y_{ij} is the dry weight of a watermellon plant grown in Row i and Column j.

α_k is the effect of the kth Treatment on dry weight

β_i is the effect of the ith Row on dry weight

γ_j is the effect of the jth Column on dry weight

b. The Row, Column, and Treatment Means are given here:

Level	1	2	3	4
Row Mean $\bar{Y}_{i.}$	1.53	1.5475	1.545	1.5475
Column Mean $\bar{Y}_{.j}$	1.5625	1.575	1.505	1.5275
Treatment Mean $\bar{Y}_k$	1.7375	1.685	1.4225	1.325

The overall mean is $\bar{Y}.. = 1.5425$

The parameter estimates are given here:

$\hat{\mu} = 1.5425$, $\hat{\beta}_1 = -.0125$, $\hat{\beta}_2 = .005$, $\hat{\beta}_3 = 0.0025$, $\beta_4 = .005$

$\hat{\gamma}_1 = .02$, $\hat{\gamma}_2 = .0325$ $\hat{\gamma}_3 = -.0375$, $\hat{\alpha}_4 = -.015$

$\hat{\alpha}_1 = .195$, $\hat{\alpha}_2 = .1425$ $\hat{\alpha}_3 = -.12$, $\hat{\alpha}_4 = -.2175$

$F = \frac{SS_{Trt}/df_{Trt}}{SS_{Error}/df_{Error}} = \frac{.48015/3}{.00075/6} = 1280.4$ with $df - 3, 6 \Rightarrow p - value < 0.0001$

Reject $H_o : \mu_1 = \mu_2 - \mu_3 = \mu_4$ and conclude there is significant evidence that the four treatments have different mean dry weights.

15.5 Factorial Treatment Structure in a CRD

15.12 a. Ten reps of a completely randomized design with a 3x2 factorial treatment structure.

b. $y_{ijk} = \mu + \alpha_i + \beta_j + \alpha\beta_{ij} + \epsilon_{ijk}$; $i = 1, 2, 3$; $j = 1, 2$; $k = 1, \cdots, 10$;

where y_{ijk} is the attention span of the kth child of Age i viewing Product j.

α_i is the effect of the ith Age on attention span

β_j is the effect of the jth Product on attention span

$\alpha\beta_{ij}$ is the interaction effect of the ith Age and jth Product on attention span

c. The treatment means are given here:

A_1P_1	A_2P_1	A_3P_1	A_1P_2	A_2P_2	A_3P_2
22.9	19.6	21.9	23.1	30.5	45.6
A_1	A_2	A_3		P_1	P_2
23.0	25.05	33.75		21.47	33.07

$$\hat{\mu} = 27.27, \quad \hat{\alpha}_1 = -4.27, \quad \hat{\alpha}_2 = -2.22, \quad \hat{\alpha}_3 = 6.48$$
$$\hat{\beta}_1 = -5.8, \quad \hat{\beta}_2 = 5.8$$
$$\hat{\alpha\beta}_{11} = 5.7, \quad \hat{\alpha\beta}_{21} = .35, \quad \hat{\alpha\beta}_{31} = -6.05$$
$$\hat{\alpha\beta}_{12} = -5.7, \quad \hat{\alpha\beta}_{22} = -.35, \quad \hat{\alpha\beta}_{32} = 6.05$$

d. The ANOVA table is given here:

Source	DF	SS	MS	F	p-value
Age	2	1303.0	651.5	4.43	0.017
Product	1	2018.4	2018.4	13.72	0.001
Interaction	2	1384.3	692.1	4.70	0.013
Error	54	7944	147.1		
Total	59	12649.7			

15.13   a. A profile plot of the data is given here:

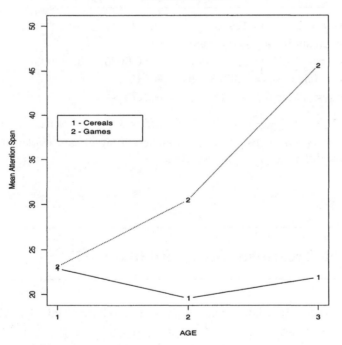

The profile plot indicates an increasing effect of Product Type as Age increases.

b. The p-value for the interaction term is 0.013. There is significant evidence of an interaction between the factors Age and Product Type. Thus, the amount of difference in mean attention of children between breakfast cereals and video games would vary across the three age groups. From the profile plots, the estimated mean attention span for video games is larger than for breakfast cereals, with the size of the difference becoming larger as age increases.

160

15.15 There are 15 treatments consisting of the 3 levels of Factor A combined with the 5 levels of Factor B. These 15 treatments will be randomly assigned to 15 experimental units in each of the three blocks as seen in the following diagram:

	Block 1			Block 2			Block 3		
	Factor B			Factor B			Factor B		
Factor A	B1	B2	B3	B1	B2	B3	B1	B2	B3
A1	x	x	x	x	x	x	x	x	x
A2	x	x	x	x	x	x	x	x	x
A3	x	x	x	x	x	x	x	x	x
A4	x	x	x	x	x	x	x	x	x
A5	x	x	x	x	x	x	x	x	x

Source	DF	SS	MS	F	p-value
Treatment	14	SST	MST		
Factor A	2	SSA	MSA	MSA/MSE	
Factor B	4	SSB	MSB	MSB/MSE	
Interaction	8	SSAB	MSAB	MSAB/MSE	
Blocks	2	SSBL	MSBL		
Error	28	SSE	MSE		
Total	44	SSTOT			

Supplementary Exercises

15.17 A randomized complete block design with days as blocks and treatments consisting of the 3x4 temperature-pressure combinations. The twelve treatments would be randomly assigned to twelve samples on each of the three days so that one replication of the 3x4 factorial experiment would be observed each day. The order in which the 12 bonding strength readings are taken should be randomized each day. A diagram is given here:

	Day 1			Day 2			Day 3		
	Temperature			Temperature			Temperature		
Pressure	$280°F$	$300°F$	$320°F$	$280°F$	$300°F$	$320°F$	$280°F$	$300°F$	$320°F$
100	S6	S1	S12	S9	S5	S1	S12	S3	S7
150	S3	S11	S8	S6	S4	S10	S8	S9	S2
200	S5	S7	S4	S2	S3	S7	S4	S5	S11
250	S10	S9	S2	S11	S8	S12	S6	S10	S1

15.21 a. The design is a completely randomized 4x4 factorial experiment with Factor A - Cu Rate and Factor B - Mn Rate. There are two replications of the 16 treatments.

b. A model for this experiment is given here:

$$y_{ijk} = \mu + \alpha_i + \beta_j + \alpha\beta_{ij} + \epsilon_{ijk}; \quad i = 1,2,3,4; \quad j = 1,2,3,4; \quad k = 1,2;$$

where y_{ijk} is the soybean yield of the kth plot using the ith Cu Rate and jth Mn Rate.

α_i is the effect of the ith Cu Rate on yield

β_j is the effect of the jth Mn Rate on yield

$\alpha\beta_{ij}$ is the interaction effect of the ith Cu Rate and jth Mn Rate on soybean yield.

c. A profile plot of the data is given here:

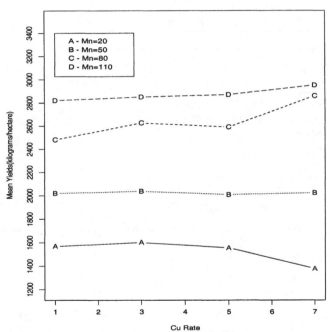

Based on the profile plot, there appears to be a strong interaction between the factors Cu Rate and Mn Rate. The mean soybean yield increases for increasing Cu Rate at a Mn Rate of 80 but the mean soybean yield stays constant initially and then decreases for increasing Cu Rate at a Mn Rate of 20. At a Mn Rate of 110 and 50, the mean soybean yield remains relatively constant with increasing rates of Cu.

15.22 a. The test for an interaction has F=11.34 with df=9,16 which yields a p-value=0.0001. This implies there is significant evidence of an interaction between Cu Rate and Mn Rate on Soybean yield.

b. Mn = 110

162

c. Cu = 7

d. (Cu,Mn)=(7,110)

15.28   a. The design is a completely randomized 3x9 factorial experiment with five replications; Factor A is Level of Severity and Factor B is Type of Medication.

b. A model for this experiment is given here:

$y_{ijk} = \mu + \alpha_i + \beta_j + \alpha\beta_{ij} + \epsilon_{ijk}; \quad i = 1,2,3; \quad j = 1,\cdots,9; \quad k = 1,2,3,4,5;$

where y_{ijk} is the temperature of the kth patient having the ith severity level using the jth medication:

α_i is the effect of the ith severity level on temperature

β_j is the effect of the jth medication on temperature

$\alpha\beta_{ij}$ is the interaction effect of the ith severity level and jth medication on temperature

15.29   a. The complete AOV table is given here:

Source	DF	SS	MS	F	p-value
Severity	2	0.3628	0.1814	7.37	0.0010
Medication	8	3.5117	0.4390	17.85	0.0001
Interaction	16	0.5012	0.0313	1.27	0.2297
Error	108	2.6520	0.0246		
Total	134	7.0277			

b. Yes, because the p-value$-$.2297 for the test of an interaction which indicates that the interaction between medication and severity is not significant. Therefore, differences in the nine medications should be consistent for the three levels of severity.

c. Medication: $p - value < 0.0001 \Rightarrow$ There is significant evidence of a difference in the mean temperatures for the nine medications.

Severity: $p - value = 0.0010 \Rightarrow$ There is significant evidence of a difference in the mean temperatures for the three levels of severity.

d. The profile plot is given here:

Profile Plot

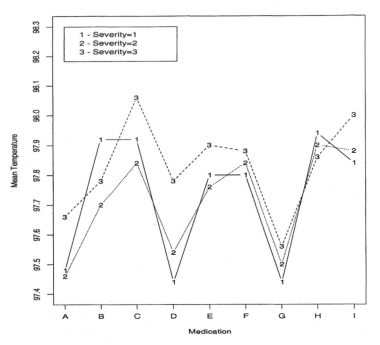

The profile plot confirms our conclusion that the three levels of severity have a consistent pattern across the nine medications after taking into account the variation in the sample means as estimators of the population means, i.e., the size of $SE(\bar{Y}_{ij.}) = \sqrt{.0246/5} = .07$.

15.33 a. The effect of grove is completely confounded with the pH effect. It would be impossible to determine if the main effects of pH and the interaction between pH and Ca Rate are due to pH differences or differences in the four groves.

b. The grove factor would be treated as a blocking factor that takes into account differences in temperature, rainfall, pesticide usage, air pollution, other soil characteristics besides pH, and other conditions in the grove that may impact growth. Plots of land within a grove are more likely to be similar with respect to these characteristics than plots in different groves. By controlling the pH of the soil by application of selected chemicals, identify 12 plots within each grove, three of each level of pH. Randomly assign a rate of Ca to plots of each level of pH as demonstrated here:

pH	Grove 1			Grove 2			Grove 3			Grove 4		
	100	200	300	100	200	300	100	200	300	100	200	300
4.0	P3	P5	P6	P12	P10	P6	P3	P8	P11	P7	P4	P10
5.0	P10	P8	P12	P3	P2	P11	P9	P4	P5	P2	P5	P11
6.0	P1	P11	P9	P7	P4	P1	P10	P6	P7	P1	P8	P9
7.0	P7	P2	P4	P9	P8	P5	P12	P2	P1	P3	P6	P12

15.35 a. A randomized complete block design with a 3x2x2 factorial treatment structure. The blocks are the six panels, Factor A is Sweetness, Factor B is Caloric Content, and Factor C is Color. There is one replication of the complete experiment.

b. A model for this experiment is given here:

$y_{ijkm} = \mu + \nu_i + \alpha_j + \beta_k + \alpha\beta_{jk} + \gamma_m + \alpha\gamma_{jm} + \beta\gamma_{km} + \alpha\beta\gamma_{jkm} + \epsilon_{ijkm};\ i = 1, \cdots, 6;\ j = 1, 2, 3;\ k = 1, 2;\ m = 1, 2;$

where y_{ijkm} is the rating of the ith panel of a drink formulated with the jth sweetness level, kth caloric level, and mth color:

ν_i is the effect of the ith panel on rating

α_j is the effect of the jth sweetness level on rating

β_k is the effect of the kth caloric level on rating

$\alpha\beta_{jk}$ is the interaction effect of the jth sweetness level and kth caloric level on rating

γ_m is the effect of the mth color on rating

$\alpha\gamma_{jm}$ is the interaction effect of the jth sweetness level and mth color on rating

$\beta\gamma_{km}$ is the interaction effect of the kth caloric level and mth color on rating

$\alpha\beta\gamma_{jkm}$ is the interaction effect of the jth sweetness level, kth caloric level and mth color on rating

c. The complete AOV table is given here:

Source	DF	SS	MS	F	p-value
Panels	5	SSP	SSP/5		
Treatments	11	SST	SST/11	MST/MSE	
Sweetness(A)	2	SSA	SSA/2	MSA/MSE	
Caloric(B)	1	SSB	SSB/1	MSB/MSE	
SSAB	2	SSAB	SSAB/2	MSAB/MSE	
Color(C)	1	SSC	SSC/1	MSC/MSE	
SSAC	2	SSAC	SSAC/2	MSAC/MSE	
SSBC	1	SSBC	SSBC/1	MSBC/MSE	
SSABC	2	SSABC	SSABC/2	MSABC/MSE	
Error	55	SSE	SSE/55		
Total	71	SST			

15.37 The three-way interaction has p-value=0.6459 and hence is not significant. The two-way interaction between Caloric level and Color has p-value=0.0927 and hence is not significant. The two-way interactions between Sweetness level and Caloric level has p-value=0.0307 and the two-way interaction between Sweetness level and Color has p-value=0.0004. Thus, Sweetness has a significant interaction with both Caloric level and Color. To determine the effect of Sweetness on the ratings it would be necessary to compare the three levels of Sweetness separately at the two levels of Caloric level and then separately at the two levels of Color. These differences could be displayed through a profile plot of Sweetness by Caloric Content and a profile plot of Sweetness by Color. Furthermore, we could conduct a Tukey comparison of the mean ratings between the levels of Sweetness across the two levels of Caloric level and then across the two Colors. The plots and Tukey analysis is given here:

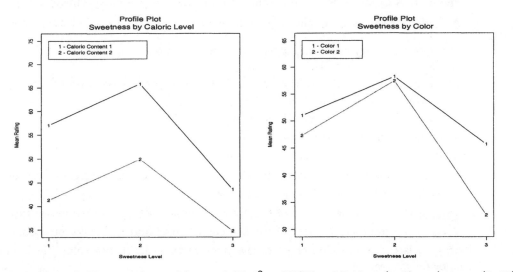

Using Tukey's W-procedure with $\alpha = 0.05$, $s_\epsilon^2 = MSE = 27.48$, $q_\alpha(t, df_{error}) = q_{.05}(3, 60) = 3.40 \Rightarrow$

$$W = (3.40)\sqrt{\frac{27.48}{12}} = 5.15 \Rightarrow$$

		Sweetness Level		
Caloric Content		1	2	3
1	Mean	57.00	65.75	43.55
	Grouping	b	c	a
2	Mean	41.30	49.85	34.75
	Grouping	b	c	a
		Sweetness Level		
Color		1	2	3
1	Mean	51.00	58.20	45.65
	Grouping	b	c	a
2	Mean	47.30	57.40	32.65
	Grouping	b	c	a

From the above table we observe that the interactions although significant were all orderly. Thus, the mean ratings were in the same order for the three levels of Sweetness for the two levels of Caloric content and two colors. The Sweetness level of 3 produced the lowest mean ratings and Sweetness level of 2 produced the highest mean ratings for all levels of Color and Caloric content.

15.40   a. The experiment is run as three reps of a completely randomized design with a 2x4 factorial treatment structure. A model for the experiment is given here:

$$y_{ijk} = \mu + \alpha_i + \beta_j + \alpha\beta_{ij} + \epsilon_{ijk}; \quad i = 1,2,3,4; \quad j = 1,2; \quad k = 1,2,3;$$

where y_{ijk} is the amount of active ingredient (or pH) of the kth vial having the ith storage time in laboratory jth:

α_i is the effect of the *ith* storage time on amount of active ingredient (or pH)

β_j is the effect of the *jth* laboratory on amount of active ingredient (or pH)

$\alpha\beta_{ij}$ is the interaction effect of the *ith* storage time and *jth* laboratory on amount of active ingredient (or pH)

b. The complete AOV table is given here:

Source	DF	SS	MS	F	p-value
Storage Time	3	SSA	SSA/3	MSA/MSE	
Laboratory	1	SSB	SSB/1	MSB/MSE	
Interaction	3	SSAB	SSAB/3	MSAB/MSE	
Error	16	SSE	SSE/16		
Total	23	SST			

15.41 The analysis for Amount of Active Ingredient:

There is significant evidence (p-value=0.0003) of an interaction between laboratory and time in storage. Therefore, a Tukey analysis of storage time will be conducted separately for each laboratory. The profile plot and Tukey analysis is given here:

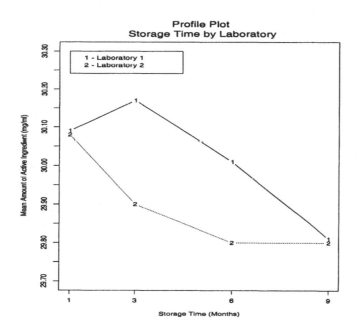

Using Tukey's W-procedure with $\alpha = 0.05, s_\epsilon^2 = MSE = 0.0024458, q_\alpha(t, df_{error}) = q_{.05}(4, 16) = 4.05 \Rightarrow$

$W = (4.05)\sqrt{\frac{.0024458}{3}} = 0.116 \Rightarrow$

		Storage Time			
Laboratory		1	3	6	9
1	Mean	30.09	30.17	30.01	29.81
	Grouping	bc	c	b	a
2	Mean	30.08	29.90	29.80	29.80
	Grouping	b	a	a	a

From the above table we observe that for Laboratory 1, the mean amount of the active ingredient was lowest after a storage time of 9 months. There was not a significant difference in the mean amounts of the active ingredient for 1 and 3 months of storage and for 1 and 6 months of storage. For Laboratory 2, the results were somewhat different. There was a significant decline in the mean amount of the active ingredient after the first month of storage but the mean did not significantly change during the 3, 6, and 9 months of storage.

The analysis for pH:

There is not significant evidence (p-value=0.4038) of an interaction between laboratory and time in storage. Therefore, a Tukey analysis will be conducted for storage time and for laboratory. The profile plot and Tukey analysis is given here:

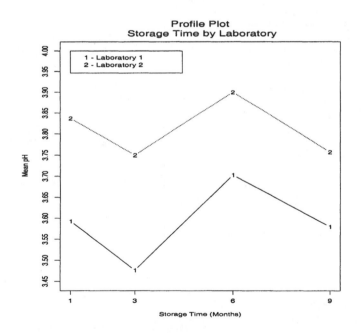

Using Tukey's W-procedure with $\alpha = 0.05$, $s_\epsilon^2 = MSE = .0027958$, $q_\alpha(t, df_{error}) = q_{.05}(3, 16)$
$3.65 \Rightarrow$

$$W = (3.65)\sqrt{\frac{.0027958}{6}} = .0788 \Rightarrow$$

	Storage Time			
	1	3	6	9
Mean	3.715	3.614	3.802	3.669
Grouping	b	a	c	ab

The mean pH is lowest at storage times of 3 and 9 months and highest at 6 months.

The p-value for testing a difference in the mean pH between the two laboratories was less than 0.0001. Thus, there significant evidence that the vials have lower mean pH values at Laboratory 1 than at Laboratory 2 (3.588 vs 3.811).

15.42   a. With respect to Amount of Active Ingredient:

The profile plots for the two-way interactions are given here:

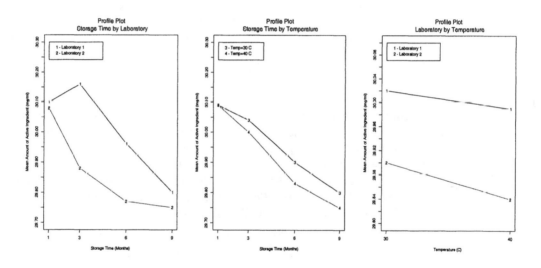

The Time*Lab*Temp interaction (p-value=.6914), Lab*Temp interaction (p-value=.3519), and Time*Temp interaction (p-value=.2817) were not significant. However, there was significant evidence (p-value=.0192) of a difference in the means for Amount of Active Ingredient at the two temperatures. There is a strong interaction between Time and Lab (p-value<.0001). Thus, comparisons of the means at the four storage times should be done separately for each Lab. The main effects of Lab and Time are not informative since these two factors have a significant interaction.

b. With respect to pH:

The profile plots for the two-way interactions are given here:

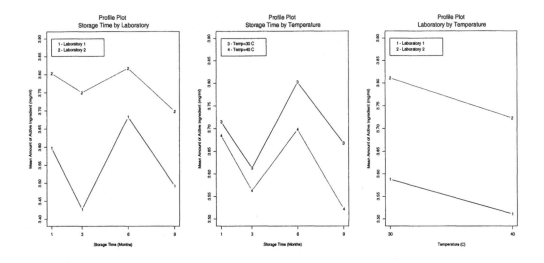

The Time*Lab*Temp interaction (p-value=.0621), Lab*Temp interaction, p-value = .7617, and Time*Temp interaction, p-value=.0686, were not significant. However, there was significant evidence, p-value < .0001, of a difference in the means for pH at the two temperatures. There is a strong interaction between Time and Lab, p-value=.0028. Thus, comparisons of the means at the four storage times should be done separately for each Lab. The main effects of Lab and Time are not informative since these two factors have a significant interaction.

c. Because the interaction between storage time and laboratory is significant, the effects of storage time differ between the two labs. Lab 1 has highest mean pH after 6 months, followed by months 1,9 and 3 months, respectively. In contrast, Lab 2 shows the highest mean pH for 6 months and 1 month, followed by 3 and 9 months, respectively.

Because the interaction between temperature and laboratory is not significant, the effects of temperature are the same for the two labs, with highest pH at 30°C.

d. Because the interaction between storage time and laboratory is significant, the effects of storage time differ between the two labs. Lab 1 has the highest mean active ingredient concentration after 3 months of storage time, followed by 1, 6, and 9 months, respectively. In contrast, Lab 2 shows its highest mean active ingredient concentration after 1 month of storage time, followed by 3, 6, and 9 months, respectively.

Because the interaction between temperature and laboratory is not significant, the effects of storage time are the same for the two labs, with highest pH at 30°C..

15.43   a. The test for an interaction yields p-value=0.0255. There is significant evidence that an interaction exists between Ratio and Supply in regards to the mean Profit. The following profile plot displays the interaction:

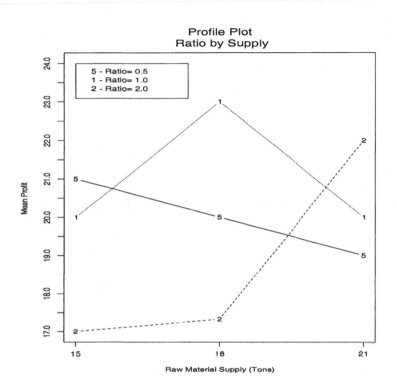

b. Because of the significant interaction between the factors Ratio and Supply, it is not possible to consider the factors separately. Therefore, the nine treatments consisting of nine different combinations of Ratio and Supply will be examined using LSD-procedure with $\alpha = 0.05, s_\epsilon^2 = MSE = 4.592593, t_{.025,18} = 2.101 \Rightarrow$

$$LSD = (2.101)\sqrt{\frac{(2)(4.592593)}{3}} = 3.68 \Rightarrow$$

	Ratio-Supply								
	2:15	2:18	.5:21	.5:18	1:15	1:21	.5:15	2:21	1:18
Mean	17	17.33	19	20	20	20	21	22	23
Grouping	a	ab	abc	abcd	abcd	abcd	bcd	cd	d

The combinations of Ratio and Supply yielding the highest mean profits are (ratio=.5,Supply=15), (Ratio=.5,Supply=18), (Ratio=1,Supply=15), (Ratio=1,Supply=18), (Ratio=1,Supply=21), (Ratio=2,Supply=21).

These six combinations do not have significantly different mean profits.

15.45 a. Randomized Complete Block Design with the 5 specimens of fabrics serving as the blocks and the three Dyes being the treatments.

b. The test for the differences in mean quality of the three dyes has p-value=0.0100. Thus, there is significant evidence of a difference in the mean quality of the three dyes.

Using Tukey's W-procedure with $\alpha = 0.05, s_\epsilon^2 = MSE = 34.367, q_\alpha(t, df_{error}) = q_{.05}(3, 8) = 4.04 \Rightarrow$

$$W = (4.04)\sqrt{\frac{34.367}{5}} = 10.59 \Rightarrow$$

	Dye		
	A	B	C
Mean	77.40	84.60	92.80
Grouping	a	ab	b

c. $t = 3, b = 5 \Rightarrow RE(RCB, CR) = \frac{(5-1)(23.567)+(5)(3-1)(34.367)}{((5)(3)-1)(34.367)} = 0.91 \Rightarrow$

It would take 0.91 times as many observations (approximately 5) per treatment in a completely randomized design to achieve the same level of precision in estimating the treatment means as was accomplished in the randomized complete block design. Since RE was slightly less than 1, we would conclude that the blocking was not effective.

15.47   a. Latin Square Design with blocking variables Farm and Fertility. The treatment is the five types of fertilizers.

   b. There is significant evidence $(p - value < 0.0001)$ the mean yields are different for the five fertilizers.

15.48 Using Tukey's W-procedure with $\alpha = 0.05, s_\epsilon^2 = MSE = .3886, q_\alpha(t, df_{error}) = q_{.05}(5, 12) = 4.52 \Rightarrow$

$$W = (4.52)\sqrt{\frac{.3886}{5}} = 1.26 \Rightarrow$$

	Fertilizer				
	A	B	C	D	E
Mean	5.32	6.56	7.64	7.88	8.24
Grouping	a	ab	bc	c	c

The following pairs of fertilizers have significantly different mean yields:

(A,C), (A,D), (A,E), (B,D), (B,E)

Chapter 16: The Analysis of Covariance

16.2 A Completely Randomized Design with One Covariate

16.1 $y_i = \beta_o + \beta_1 x_{1i} + \beta_2 x_{2i} + \beta_3 x_{3i} + \beta_4 x_{4i} + \beta_5 x_{5i}$
$\qquad + \beta_6 x_{1i} x_{2i} + \beta_7 x_{1i} x_{3i} + \beta_8 x_{1i} x_{4i} + \beta_9 x_{1i} x_{5i} + \epsilon_i \quad$ for $\quad i = 1, \cdots, 30$

$x_1 = $ Covariate

$x_2 = \begin{cases} 1 & \text{if} \quad \text{Treatment 2 is applied} \\ 0 & \text{if} \quad \text{Otherwise} \end{cases} \qquad x_3 = \begin{cases} 1 & \text{if} \quad \text{Treatment 3 is applied} \\ 0 & \text{if} \quad \text{Otherwise} \end{cases}$

$x_4 = \begin{cases} 1 & \text{if} \quad \text{Treatment 4 is applied} \\ 0 & \text{if} \quad \text{Otherwise} \end{cases} \qquad x_5 = \begin{cases} 1 & \text{if} \quad \text{Treatment 5 is applied} \\ 0 & \text{if} \quad \text{Otherwise} \end{cases}$

The coefficients are identified through the mean response under each treatment:

Treatment	Mean Response
1	$\beta_o + \beta_1 x_1$
2	$(\beta_o + \beta_2) + (\beta_1 + \beta_6)x_1$
3	$(\beta_o + \beta_3) + (\beta_1 + \beta_7)x_1$
4	$(\beta_o + \beta_4) + (\beta_1 + \beta_8)x_1$
5	$(\beta_o + \beta_5) + (\beta_1 + \beta_9)x_1$

16.3 Parallelism results from the five lines having the same slope, i.e., the correct model follows the null hypothesis:

$H_o : \beta_6 = \beta_7 = \beta_8 = \beta_9 = 0.$

A test of H_o versus the alternative

$H_a :$ At least one of $\beta_6, \beta_7, \beta_8, \beta_9$ is not zero

consists of fitting two models:

Model 1 (Assumes H_a is true): $y_i = \beta_o + \beta_1 x_{1i} + \beta_2 x_{2i} + \beta_3 x_{3i} + \beta_4 x_{4i} + \beta_5 x_{5i} + \beta_6 x_{1i} x_{2i} + \beta_7 x_{1i} x_{3i} + \beta_8 x_{1i} x_{4i} + \beta_9 x_{1i} x_{5i} + \epsilon$

Model 2 (Assumes H_o is true): $y_i = \beta_o + \beta_1 x_{1i} + \beta_2 x_{2i} + \beta_3 x_{3i} + \beta_4 x_{4i} + \beta_5 x_{5i} + \epsilon$

Let $MS_{drop} = (SSE_2 - SSE_1)/(df_{E2} - df_{E1}) = (SSE_2 - SSE_1)/((30 - 6) - (30 - 10))$

The test statistic is $F = \frac{MS_{drop}}{MSE_1}$, with $df_1 = (30 - 6) - (30 - 10) = 4, \quad df_2 = 30 - 10 = 20.$

16.5 A scatterplot of the data is given here:

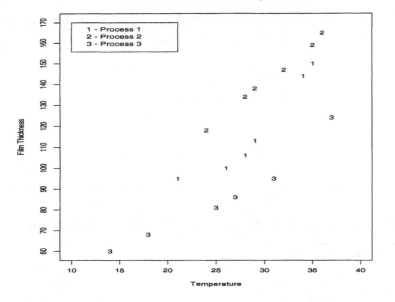

Film Thickness vs Temperature

The analysis of variance tables for three models are given here: $x =$ Temperature

$$I_1 = \begin{cases} 1 & \text{if} & \text{Process 2 is applied} \\ 0 & \text{if} & \text{Otherwise} \end{cases} \qquad I_2 = \begin{cases} 1 & \text{if} & \text{Process 3 is applied} \\ 0 & \text{if} & \text{Otherwise} \end{cases}$$

Model 1 (unequal slopes): $y_i = \beta_o + \beta_1 x_i + \beta_2 I_{1i} + \beta_3 I_{2i} + \beta_4 x_i I_{1i} + \beta_5 x_i I_{2i} + \epsilon_i$

Model 2 (equal slopes): $y_i = \beta_o + \beta_1 x_i + \beta_2 I_{1i} + \beta_3 I_{2i} + \epsilon_i$

Model 3 (equal slopes and intercepts): $y_i = \beta_o + \beta_1 x_i + \epsilon_i$

Model 1:

Regression Analysis: y versus I1, I2, x, x*I1, x*I2
Analysis of Variance

Source	DF	SS	MS	F	P
Regression	5	16396.2	3279.2	89.15	0.000
Residual Error	12	441.4	36.8		
Total	17	16837.6			

Model 2:

Regression Analysis: y versus I1, I2, x

The regression equation is
y = 26.081 + 19.655 I1 - 21.176 I2 + 3.1879 x

Analysis of Variance

Source	DF	SS	MS	F	P
Regression	3	16092.2	5364.1	100.74	0.000
Residual Error	14	745.4	53.2		
Total	17	16837.6			

Model 3:

Regression Analysis: y versus x

The regression equation is
y = - 1.71 + 4.1529 x

Analysis of Variance

Source	DF	SS	MS	F	P
Regression	1	11721	11721	36.65	0.000
Residual Error	16	5116	320		
Total	17	16838			

The test of equal slopes:

$MS_{drop} = (745.4 - 441.4)/(14 - 12) = 152 \Rightarrow F = \frac{152}{36.8} = 4.13$ with $df = 2, 12 \Rightarrow$

$p - value = Pr(F_{2,12} \geq 4.13) = 0.0432 > 0.01 \Rightarrow$

Fail to reject H_o and conclude that there is not significant evidence that the slopes are different.

The test of difference in mean film thickness for the three processes:

$MS_{drop} = (5116 - 745.4)/(16 - 14) = 2185.3 \Rightarrow F = \frac{2185.3}{53.2} = 41.08$ with $df = 2, 14 \Rightarrow$

$p - value = Pr(F_{2,14} \geq 41.08) = 0.0001 < 0.01 \Rightarrow$

Reject H_o and conclude that there is significant evidence that the adjusted process means are different.

16.4 Multiple Covariates and More Complicated Designs

16.6 $y_i = \beta_o + \beta_1 x_{1i} + \beta_2 x_{2i} + \beta_3 x_{3i} + \beta_4 x_{4i} + \beta_5 x_{1i} x_{2i} + \beta_6 x_{1i} x_{3i} + \beta_7 x_{1i} x_{4i}$

$\qquad + \beta_8 x_{5i} + \beta_9 x_{6i} + \beta_{10} x_{7i} + \beta_{11} x_{8i} + \beta_{12} x_{9i} + \beta_{13} x_{10i} + \epsilon_i \quad$ for $\ i = 1, \cdots, 16$

$x_1 = $ Covariate

$$x_2 = \begin{cases} 1 & \text{if} & \text{Treatment 2 is applied} \\ 0 & \text{if} & \text{Otherwise} \end{cases} \qquad x_3 = \begin{cases} 1 & \text{if} & \text{Treatment 3 is applied} \\ 0 & \text{if} & \text{Otherwise} \end{cases}$$

$$x_4 = \begin{cases} 1 & \text{if} & \text{Treatment 4 is applied} \\ 0 & \text{if} & \text{Otherwise} \end{cases} \qquad x_5 = \begin{cases} 1 & \text{if} & \text{Observation in Row 2} \\ 0 & \text{if} & \text{Otherwise} \end{cases}$$

$$x_6 = \begin{cases} 1 & \text{if} & \text{Observation in Row 3} \\ 0 & \text{if} & \text{Otherwise} \end{cases} \qquad x_7 = \begin{cases} 1 & \text{if} & \text{Observation in Row 4} \\ 0 & \text{if} & \text{Otherwise} \end{cases}$$

$$x_8 = \begin{cases} 1 & \text{if} & \text{Observation in Column 2} \\ 0 & \text{if} & \text{Otherwise} \end{cases} \qquad x_9 = \begin{cases} 1 & \text{if} & \text{Observation in Column 3} \\ 0 & \text{if} & \text{Otherwise} \end{cases}$$

$$x_{10} = \begin{cases} 1 & \text{if} & \text{Observation in Column 4} \\ 0 & \text{if} & \text{Otherwise} \end{cases}$$

16.7 a. Obtain SSE_1 from the model in Exercise 16.6. Obtain SSE_2 from the model assuming the lines are parallel: $y_i = \beta_o + \beta_1 x_{1i} + \beta_2 x_{2i} + \beta_3 x_{3i} + \beta_4 x_{4i} + \beta_8 x_{5i} + \beta_9 x_{6i} + \beta_{10} x_{7i} + \beta_{11} x_{8i} + \beta_{12} x_{9i} + \beta_{13} x_{10i} + \epsilon_i$

$MS_{drop} = (SSE_2 - SSE_1)/(5-2)$

$F = \frac{MS_{drop}}{MSE_1}$, with df=3,2

b. Obtain SSE_1 from the model assuming the lines are parallel:

$y_i = \beta_o + \beta_1 x_{1i} + \beta_2 x_{2i} + \beta_3 x_{3i} + \beta_4 x_{4i} + \beta_8 x_{5i} + \beta_9 x_{6i} + \beta_{10} x_{7i} + \beta_{11} x_{8i} + \beta_{12} x_{9i} + \beta_{13} x_{10i} + \epsilon_i$

Obtain SSE_2 from the model assuming no treatment difference:

$y_i = \beta_o + \beta_1 x_{1i} + \beta_8 x_{5i} + \beta_9 x_{6i} + \beta_{10} x_{7i} + \beta_{11} x_{8i} + \beta_{12} x_{9i} + \beta_{13} x_{10i} + \epsilon_i$

$MS_{drop} = (SSE_2 - SSE_1)/(5-2)$

$F = \frac{MS_{drop}}{MSE_1}$, with df=3,5

Supplementary Exercises

16.9 a. Randomized complete block design with the three antidepressants as treatments, age-sex combinations as six blocks, and the pretreatment rating serving as a covariate.

b. $y_i = \beta_o + \beta_1 x_{1i} + \beta_2 x_{2i} + \beta_3 x_{3i} + \beta_4 x_{1i} x_{2i} + \beta_5 x_{1i} x_{3i}$
$\qquad + \beta_6 x_{4i} + \beta_7 x_{5i} + \beta_8 x_{6i} + \beta_9 x_{7i} + \beta_{10} x_{8i} + \epsilon_i \quad \text{for} \quad i = 1, \cdots, 16$

$x_1 = $ Covariate

$$x_2 = \begin{cases} 1 & \text{if} & \text{Antidepressant B} \\ 0 & \text{if} & \text{Otherwise} \end{cases} \qquad x_3 = \begin{cases} 1 & \text{if} & \text{Antidepressant C} \\ 0 & \text{if} & \text{Otherwise} \end{cases}$$

$$x_4 = \begin{cases} 1 & \text{if} & \text{Observation in Block 2} \\ 0 & \text{if} & \text{Otherwise} \end{cases} \qquad x_5 = \begin{cases} 1 & \text{if} & \text{Observation in Block 3} \\ 0 & \text{if} & \text{Otherwise} \end{cases}$$

$$x_6 = \begin{cases} 1 & \text{if} & \text{Observation in Block 4} \\ 0 & \text{if} & \text{Otherwise} \end{cases} \qquad x_7 = \begin{cases} 1 & \text{if} & \text{Observation in Block 5} \\ 0 & \text{if} & \text{Otherwise} \end{cases}$$

$$x_8 = \begin{cases} 1 & \text{if} & \text{Observation in Block 6} \\ 0 & \text{if} & \text{Otherwise} \end{cases}$$

16.10   a. Minitab output for the three models is given here:

```
Model 1: Unequal Slopes

General Linear Model: y versus Block, Antidepressant


Factor     Type Levels Values
Block      fixed    6 1 2 3 4 5 6
Antidepr   fixed    3 A         B         C

Analysis of Variance for y, using Adjusted SS for Tests

Source      DF   Adj SS    Adj MS     F     P
x            1    7.079     7.079   0.77  0.408
Block        5    9.896     1.979   0.22  0.944
Antidepr     2    2.023     1.012   0.11  0.897
Antidepr*x   2    4.574     2.287   0.25  0.785
Error        7   64.007     9.144
Total       17  168.500


-----------------------------------------------------------------


Model 2: Parallel Lines


General Linear Model: y versus Block, Antidepressant


Factor     Type Levels Values
Block      fixed    6 1 2 3 4 5 6
Antidepr   fixed    3 A         B         C

Analysis of Variance for y, using Adjusted SS for Tests

Source      DF    Adj SS    Adj MS     F     P
x            1     3.752     3.752   0.49  0.501
Block        5     6.636     1.327   0.17  0.966
Antidepr     2    72.077    36.039   4.73  0.039
Error        9    68.581     7.620
Total       17


----------------------------------------------------------------
Model 3: No Treatment Effect

General Linear Model: y versus Block


Factor     Type Levels Values
Block      fixed    6 1 2 3 4 5 6

Analysis of Variance for y, using Adjusted SS for Tests

Source      DF      Adj SS    Adj MS     F     P
```

```
x          1      24.68    24.68    1.93  0.192
Block      5       2.04     0.41    0.03  0.999
Error     11     140.66    12.79
Total     17
```

Model 4: Block Effect is Removed from Model:

General Linear Model: y versus Antidepressant

```
Factor     Type Levels Values
Antidepr   fixed    3 A        B        C
```

Analysis of Variance for y, using Adjusted SS for Tests

```
Source     DF    Adj SS   Adj MS      F    P
x          1     0.283    0.283    0.05  0.822
Antidepr   2    67.484   33.742    6.28  0.011
Error     14    75.217    5.373
Total     17
```

b. Test for parallelism of the three treatment lines:

$F = \frac{(68.581-64.007)/(9-7)}{64.007/7} = 0.25$, with $df = 2, 7 \Rightarrow$

$p - value = Pr(F_{2,7} \geq 0.25) = 0.7855 \Rightarrow$

There is not significant evidence that the lines are not parallel.

c. Test for difference in adjusted treatment means:

$F = \frac{(140.66-68.581)/(11-9)}{68.581/9} = 4.73$, with $df = 2, 9 \Rightarrow$

$p - value = Pr(F_{2,9} \geq 4.73) = 0.0395 \Rightarrow$

There is significant evidence that the adjusted mean ratings are different for the three types of antidepressants.

16.11   a. Test for difference in adjusted means between blocks:

$F = \frac{(75.217-68.581)/(14-9)}{68.581/9} = 0.17$, with $df = 5, 9 \Rightarrow$

$p - value = Pr(F_{5,9} \geq 0.17) = 0.9673 \Rightarrow$

There is not significant evidence that the adjusted mean ratings are different for the six blocks.

b. The sum of squares associated the block indicator variables: x_4, x_5, x_6, x_7, x_8 partition the sum of squares for blocks into five separate sum of squares, a sum of squares for each indicator variable.

c. Because there was not a significant block effect, none of the five individual sum of squares would be significant.

Chapter 17: Analysis of Variance for Some Fixed-, Random-, and Mixed-Effects Models

17.2 A Random-Effects Model

17.1   a. $y_{ij} = \mu + \alpha_i + \epsilon_{ij}$; $i = 1, 2, 3, 4, 5$; $j = 1, 2, 3, 4$

   b. Minitab Output is given here:

```
Factor      Type Levels Values
Vat         random      5 Vat 1 Vat 2 Vat 3 Vat 4 Vat 5

Analysis of Variance for Potency, using Adjusted SS for Tests

Source    DF    Seq SS     Adj SS    Adj MS       F      P
Vat        4   11.9480    11.9480    2.9870   32.53  0.000
Error     15    1.3775     1.3775    0.0918
Total     19   13.3255

Least Squares Means for Potency

Vat            Mean
Vat 1         3.375
Vat 2         2.575
Vat 3         3.425
Vat 4         4.275
Vat 5         2.025
Overall       3.135
```

   The F-test for $H_o : \sigma_\alpha^2 = 0$ versus $H_a : \sigma_\alpha^2 > 0$ has $p - value < 0.0001$. Therefore, there is significant evidence of a difference in the potency of the medication across the population of vats.

17.2   a. $\hat{\mu} = \bar{y}_{..} = 3.135$

   b. A 95% C.I. on μ is given by $\bar{y}_{..} \pm t_{\alpha/2, df_{error}} \sqrt{MST/tn} \Rightarrow$

   $3.135 \pm (2.131)\sqrt{2.9870/20} \Rightarrow 3.135 \pm 0.824 \Rightarrow (2.3, 4.0)$

17.3 Extensions of Random-Effects Models

17.3   a. $y_{ij} = \mu + \alpha_i + \beta_j + \epsilon_{ij}$; $i = 1, 2, 3$; $j = 1, 2, 3$ where
   μ is the DNA concentration mean

α_i is a random effect due to the *ith* analyst

β_j is a random effect due to the *jth* subject

ϵ_{ij} is a random effect due to all other sources but analyst and subject

b. The expected Mean Squares are given here:

Source	Expected Mean Square
Analyst	$\sigma_\epsilon^2 + 3\sigma_\alpha^2$
Subject	$\sigma_\epsilon^2 + 3\sigma_\beta^2$
Error	σ_ϵ^2

17.4 Minitab output is given here:

```
General Linear Model: DNA versus Analyst, Subject


Factor     Type Levels Values
Analyst  random      3 1 2 3
Subject  random      3 1 2 3

Analysis of Variance for DNA, using Adjusted SS for Tests

Source     DF    Seq SS    Adj SS    Adj MS       F      P
Analyst     2    0.8822    0.8822    0.4411   19.37  0.009
Subject     2   33.2356   33.2356   16.6178  729.56  0.000
Error       4    0.0911    0.0911    0.0228
Total       8   34.2089
```

The F-test for $H_o : \sigma_\alpha^2 = 0$ versus $H_a : \sigma_\alpha^2 > 0$ has $p - value = 0.009$. Therefore, there is significant evidence of a difference in the DNA concentration across the population of analysts.

The F-test for $H_o : \sigma_\beta^2 = 0$ versus $H_a : \sigma_\beta^2 > 0$ has $p - value < 0.0001$. Therefore, there is significant evidence of a difference in the DNA concentration across the population of subjects.

17.4 Mixed-Effects Models

17.6   a. $y_{ijk} = \mu + \alpha_i + \beta_j + \alpha\beta_{ij} + \epsilon_{ijk}$; $i = 1, 2, 3, 4, 5$; $j = 1, 2, 3, 4$; $k = 1, 2$ where

y_{ijk} is the number of dead ants at the *kth* mound at the *ith* location using the *jth* chemical

μ is the mean number of dead ants across all possible locations treated with the four chemicals

α_i is a random effect due to the *ith* location

β_j is a fixed effect due to the *jth* chemical

$\alpha\beta_{ij}$ is a random effect due to the interaction of the ith location and the jth chemical

ϵ_{ij} is a random effect due to all other sources but location and chemical

b. The AOV table is given here:

Source	DF	SS	MS	EMS	F	P
Location	4	3.812	0.953	$\sigma_\epsilon^2 + 8\sigma_\alpha^2$	2.75	0.057
Chemical	3	180.133	60.044	$\sigma_\epsilon^2 + 2\sigma_{\alpha\beta}^2 + 10\theta_B$	44.58	0.000
Interaction	12	16.158	1.347	$\sigma_\epsilon^2 + 2\sigma_{\alpha\beta}^2$	3.89	0.004
Error	20	6.925	0.346	σ_ϵ^2		
Total	39	207.028				

17.7 The F-test for $H_o : \sigma_{\alpha\beta}^2 = 0$ versus $H_a : \sigma_{\alpha\beta}^2 > 0$ has $p-value = 0.004$. Therefore, there is significant evidence of an interaction between Locations and Chemicals.

The F-test for $H_o : \sigma_\alpha^2 = 0$ versus $H_a : \sigma_\alpha^2 > 0$ has $p-value = 0.057$. Therefore, there is not significant evidence of an effect due to Locations.

The F-test for $H_o : \beta_1 = \cdots = \beta_4 = 0$ versus $H_a :$ at least one $\beta_i \neq 0$ has $p-value < 0.0001$. Therefore, there is significant evidence of an effect due to Chemicals.

Supplementary Exercises

17.8 For the fixed-effects model, θ_A measures the differences in a fixed number of treatment means. For the random-effects model, σ_α^2 represents the difference in treatment means for the whole population of treatments from which the t treatments used in the experiment were randomly selected.

17.9   a. Factor A fixed with 2 levels, Factor B and C random with 3 and 4 levels, respectively, and 5 replications of the 24 treatments yield the following model:

$y_{ijkl} = \mu + \alpha_i + \beta_j + \alpha\beta_{ij} + \gamma_k + \alpha\gamma_{ik} + \beta\gamma_{jk} + \alpha\beta\gamma_{ijk} + \epsilon_{ijkl}$, where

μ is the overall mean response

α_i is the fixed effect of the ith level of Factor A, with $\sum \alpha_i = 0$

β_j is the random effect of the jth level of Factor B, iid $N(0, \sigma_B^2)$ r.v.'s

$\alpha\beta_{ij}$ is the random effect of the interaction of ith level of Factor A with the jth level of Factor B, $N(0, \sigma_{AB}^2)$ r.v.'s with $\sum_i \alpha\beta_{ij} = 0$

γ_k is the random effect of the kth level of Factor C, iid $N(0, \sigma_C^2)$ r.v.'s

$\alpha\gamma_{ik}$ is the random effect of the interaction of ith level of Factor A with the kth level of Factor C, $N(0, \sigma_{AC}^2)$ r.v.'s with $\sum_i \alpha\gamma_{ik} = 0$

$\beta\gamma_{jk}$ is the random effect of the interaction of jth level of Factor B with the kth level of Factor C, iid $N(0, \sigma_{BC}^2)$

$\alpha\beta\gamma_{ijk}$ is the random effect of the interaction of ith level of Factor A with the jth level of Factor B and kth level of Factor C, $N(0, \sigma_{ABC}^2)$ r.v.'s with $\sum_i \alpha\beta\gamma_{ijk} = 0$

β_j is independent of γ_k's, $\alpha\beta_{ij}$'s, $\alpha\gamma_{ik}$'s, $\beta\gamma_{jk}$'s, $\alpha\beta\gamma_{ijk}$'s

γ_k is independent of $\alpha\beta_{ij}$'s, $\alpha\gamma_{ik}$'s, $\beta\gamma_{jk}$'s, $\alpha\beta\gamma_{ijk}$'s

$\alpha\beta_{ij}$ is independent of $\alpha\gamma_{ik}$'s, $\beta\gamma_{jk}$'s, $\alpha\beta\gamma_{ijk}$'s

$\alpha\gamma_{ik}$ is independent of $\beta\gamma_{jk}$'s, $\alpha\beta\gamma_{ijk}$'s

$\beta\gamma_{jk}$ is independent of $\alpha\beta\gamma_{ijk}$'s

ϵ_{ijkl} is random effect of all other sources, iid $N(0, \sigma_\epsilon^2)$

ϵ_{ijkl} is independent of all other r.v.'s

b. The AOV table is given here:

Source	df	Expected MS
A	1	$\sigma_\epsilon^2 + 5\sigma_{ABC}^2 + 15\sigma_{AC}^2 + 20\sigma_{AB}^2 + 60\theta_A$
B	2	$\sigma_\epsilon^2 + 10\sigma_{BC}^2 + 40\sigma_B^2$
AB	2	$\sigma_\epsilon^2 + 5\sigma_{ABC}^2 + 20\sigma_{AB}^2$
C	3	$\sigma_\epsilon^2 + 10\sigma_{BC}^2 + 30\sigma_C^2$
AC	3	$\sigma_\epsilon^2 + 5\sigma_{ABC}^2 + 15\sigma_{AC}^2$
BC	6	$\sigma_\epsilon^2 + 10\sigma_{BC}^2$
ABC	6	$\sigma_\epsilon^2 + 5\sigma_{ABC}^2$
Error	96	σ_ϵ^2

c. The F-tests are given here:

Source	F-test
A	No Exact F-test
B	MSB/MSBC
AB	MSAB/MSABC
C	MSC/MSBC
AC	MSAC/MSABC
BC	MSBC/MSE
ABC	MSABC/MSE

17.13 a. The mixed effects model is more appropriate. Researchers would be concerned about specific chemicals not a population of chemicals. They would want to determine which of the four chemicals is most effective in controlling fire ants.

b. A fixed effects model would be appropriate if the researcher was only interested in a set of specific locations, such as those with specific environmental conditions, or different levels of human activity or specific soil conditions. The fixed effects model would have both the levels of chemicals and the levels of locations used in the experiment as the only levels of interest. The levels used in the experiment would not be randomly selected from a population of levels.

17.17 a. y_{ij} is the thrust measurement made by the jth investigator on the ith mixture

$y_{ij} = \mu + \alpha_i + \beta_j + \epsilon_{ij}$

μ is the overall mean thrust of all mixtures

α_i fixed effect of the ith mixture, $\sum_i \alpha_i = 0$

β_j random effect of the jth investigator, iid $N(0, \sigma_B^2)$ r.v.'s

ϵ_{ij} random effect of all other factors, iid $N(0, \sigma_\epsilon^2)$ r.v.'s

β_j's and ϵ_{ij}'s are independent

b. The AOV table is given here:

Source	df	SS	MS	Expected MS	F-test	p-value
Mixtures	3	261260.95	87086.98	$\sigma_\epsilon^2 + 5\theta_A$	1264.73	0.0001
Investigators	4	452.50	113.13	$\sigma_\epsilon^2 + 4\sigma_B^2$	1.64	0.2280
Error	12	826.30	68.86	σ_ϵ^2		
Total	19	262539.75				

There is significant $(p - value < 0.0001)$ evidence that the mixtures produce different average thrust measurements.

17.19 a. Factor A: Time with 4 fixed levels

Factor B: Temperature with 2 fixed levels

Factor C: Laboratory with 2 random levels

The AOV table is given here for MG/ML:

183

Source	df	SS	MS	Expected MS
Time	3	0.7360	0.2453	$\sigma_\epsilon^2 + 6\sigma_{AC}^2 + 12\theta_A$
Temp	1	0.0184	0.0184	$\sigma_\epsilon^2 + 12\sigma_{BC}^2 + 24\theta_B$
Time*Temp	3	0.0121	0.004025	$\sigma_\epsilon^2 + 3\sigma_{ABC}^2 + 6\theta_{AB}$
Lab	1	0.2296	0.2296	$\sigma_\epsilon^2 + 24\sigma_C^2$
Time*Lab	3	0.1443	0.0481	$\sigma_\epsilon^2 + 6\sigma_{AC}^2$
Temp*Lab	1	0.0027	0.0027	$\sigma_\epsilon^2 + 12\sigma_{BC}^2$
Time*Temp*Lab	3	0.00445	0.001483	$\sigma_\epsilon^2 + 3\sigma_{ABC}^2$
Error	32	0.0968	0.003025	σ_ϵ^2

$F_{Time} = MSTime/MSTime*Lab = 5.10, p-value = 0.107$

$F_{Temp} = MSTemp/MSTemp*Lab = 6.82, p-value = 0.233$

$F_{Time*Temp} = MSTime*Temp/MSTime*Temp*Lab = 2.71, p-value = 0.217$

$F_{Lab} = MSLab/MSE = 75.91, p-value < 0.0001$

$F_{Lab*Time} = MSLab*Time/MSE = 15.90, p-value < 0.0001$

$F_{Lab*Temp} = MSLab*Temp/MSE = 0.89, p-value = 0.352$

$F_{Lab*Time*Temp} = MSLab*Time*Temp/MSE = 0.49, p-value = 0.691$

The tests demonstrate that only Lab and the Time by Lab interaction are significant.

c. In Exercise 15.42, we found that the effect of Time and Temp were also significant.

d. Treating Lab as a fixed effect would be appropriate if the two labs used in the study were the only Labs which can test the drug product. However, if the drug product could be tested at any of a number of labs, and the labs used in the study were randomly selected from this list of labs, then Lab should be treated as a random effect.

b. The AOV table is given here for pH:

Source	df	SS	MS	Expected MS
Time	3	0.2286	0.0762	$\sigma_\epsilon^2 + 6\sigma_{AC}^2 + 12\theta_A$
Temp	1	0.0817	0.0817	$\sigma_\epsilon^2 + 12\sigma_{BC}^2 + 24\theta_B$
Time*Temp	3	0.0251	0.008364	$\sigma_\epsilon^2 + 3\sigma_{ABC}^2 + 6\theta_{AB}$
Lab	1	0.5677	0.5677	$\sigma_\epsilon^2 + 24\sigma_C^2$
Time*Lab	3	0.0556	0.0186	$\sigma_\epsilon^2 + 6\sigma_{AC}^2$
Temp*Lab	1	0.0003	0.0003	$\sigma_\epsilon^2 + 12\sigma_{BC}^2$
Time*Temp*Lab	3	0.02597	0.008656	$\sigma_\epsilon^2 + 3\sigma_{ABC}^2$
Error	32	0.1026	0.003206	σ_ϵ^2

c. $F_{Time} = MSTime/MSTime*Lab = 4.11, p-value = 0.1380$

$F_{Temp} = MSTemp/MSTemp*Lab = 272.3, p-value = 0.0385$

$F_{Time*Temp} = MSTime*Temp/MSTime*Temp*Lab = 0.97, p-value = 0.5097$

$F_{Lab} = MSLab/MSE = 177.1, p-value < 0.0001$

$F_{Lab*Time} = MSLab*Time/MSE = 5.78, p-value = 0.0028$

$F_{Lab*Temp} = MSLab*Temp/MSE = 0.09, p-value = 0.7661$

$F_{Lab*Time*Temp} = MSLab*Time*Temp/MSE = 2.70, p-value = 0.0621$

The tests demonstrate that Temp, Lab and the Time by Lab interaction are significant.

In Exercise 15.42, we found that the effect of Time and Temp were also significant.

17.21   a. This is 5 reps of a completely randomized mixed model with

Factor A: Severity is fixed with 3 levels

Factor B: Medication is random with 9 levels

The model is $y_{ijk} = \mu + \alpha_i + \beta_j + \alpha\beta_{ij} + \epsilon_{ijk}$, where

y_{ijk} is the temperature of the kth patient of the ith severity receiving the jth medication

μ is the mean temperature over all severities and medications

α_i is the fixed effect of the ith severity, $\sum_i \alpha_i = 0$

β_j is the random effect of the jth medication, iid $N(0, \sigma_B^2)$ r.v.'s

$\alpha\beta_{ij}$ is the random interaction effect of the ith severity with the jth medication, $N(0, \sigma_{AB}^2)$ r.v.'s with $\sum_i \alpha\beta_{ij} = 0$

ϵ_{ijk} is the random effect of all other factors on temperature

β_j's are independent of $\alpha\beta_{ij}$'s and ϵ_{ijk}'s

b. The AOV table is given here:

Source	DF	SS	MS	EMS	F	P
Severity	2	0.3628	0.1814	$\sigma_\epsilon^2 + 5\sigma_{AB}^2 + 45\theta_A$	5.79	0.0128
Medication	8	3.5117	0.4390	$\sigma_\epsilon^2 + 15\sigma_B^2$	17.88	0.0001
Interaction	16	0.5012	0.03133	$\sigma_\epsilon^2 + 5\sigma_{AB}^2$	1.28	0.2231
Error	108	2.6520	0.02455	σ_ϵ^2		
Total	134	7.0277				

c. The interaction between Severity and Medication is not significant ($p - value = 0.2231$), both main effects are significant: Severity ($p - value = 0.0128$), Medication ($p - value = 0.0001$). In Exercise 15.28, the conclusions are the same except the p-value for Severity was considerably smaller in Exercise 15.28. The difference is that in this case the inferences are concerning the population of medications and not just the nine medications used in the study.

17.25 The calculations of sum of squares is given here:

$\bar{y}_{..} = \sum_{ij} y_{ij}/24 = 368.8/24 = 15.367$

$TSS = \sum_{ij}(y_{ij} - \bar{y}_{..})^2 = \sum_{ij}(y_{ij} - 368.8/24)^2 = 97.25333$

$SSL = \sum_i n_i(\bar{y}_{i.} - \bar{y}_{..})^2 = (4(16.425 - 15.367)^2 + 4(13.1 - 15.367)^2 + 4(16.95 - 15.367)^2 + 4(18.3 - 15.367)^2 + 4(13.65 - 15.367)^2 + 4(13.775 - 15.367)^2 = 91.39833$

$SSA(L) = TSS - SSL = 97.25333 - 91.35833 = 5.855$

The AOV table is given here:

Source	df	SS	MS	F	p-value
Location	5	91.3983	18.2797	56.20	0.0001
Analysis within Location	18	5.8550	0.3253		
Total	23	97.2533			

There is significant evidence ($p - value < 0.0001$) that the mean sulfur content is different for the six locations.

17.27 a. This is a randomized block split-plot design. Tasters are the blocks, Fat levels are the whole plot factor, and Method of cooking is the split-plot factor. There is a single replication of the experiment.

b. $y_{ijk} = \mu + \alpha_i + \beta_j + \alpha\beta_{ij} + \gamma_k + \gamma\alpha_{ik} + \epsilon_{ijk}$, where

y_{ijk} is the taste score from the kth taster for a meat sample having the ith fat level cooked using method j

μ is the mean taste score

α_i is fixed efffect of ith fat level

β_j is fixed effect of jth cooking method

$\alpha\beta_{ij}$ is interaction effect of ith fat level with jth cooking method

γ_k is fixed efffect of kth taster

$\gamma\alpha_{ik}$ is the whole plot random effect

ϵ_{ijk} is random effect due to all other factors

c. Note that the computer output has the wrong F-test for the main effect due to Fat level. The correct F-test has MSF/MST*F as given here:

Source	DF	SS	MS	F	P
Taster(T)	3	291.889	97.296	*	*
Fat(F)	2	146.000	73.000	0.97	0.4315
T*F	6	451.778	75.296	*	*
Method(M)	2	32.167	16.083	1.75	0.2019
F*M	4	9.833	2.458	0.27	0.8949
Error	18	165.333	9.185		
Total	35	1097.000			

d. The interaction between Method of Cooking and Level of Fat is not significant (p-value=0.8949). The main effects of Method of Cooking and Level of Fat are both nonsignificant (p-value=0.2019, p-value=0.4315, respectively). Thus, there is not significant evidence that either Method of Cooking or Level of Fat have an effect on the taste of the meat.

Chapter 18: Repeated Measures and Crossover Designs

18.3 Two-Factor Experiments with
Repeated Measures on One of the Factors

18.3 a. The mean and standard deviation of percentage inhibition by Treatment and Time are given here:

Treatment(Means)	Time				
	1	2	3	4	8
Antihistamine	20.70	28.57	31.24	29.44	25.63
Placebo	-0.76	12.55	18.23	24.79	17.57
Treatment(St.Dev.)	1	2	3	4	8
Antihistamine	23.98	12.00	14.30	12.65	14.26
Placebo	12.26	10.43	10.83	6.91	7.83

The antihistamine treated patients uniformly, across all five hours, have larger mean percentage inhibition than the placebo treated patients. The patern for the standard deviations is similar with somewhat higher values during the first hour after treatment.

b. A profile plot of the water loss data is given here:

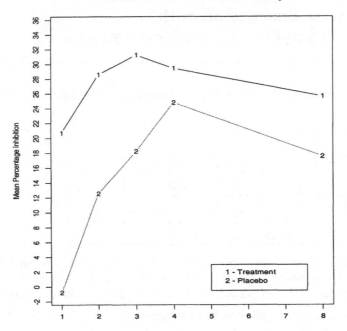

Profile Plot for Antihistamine Study

Yes, the antihistamine treated patients appear to have a higher mean percentage inhibition than the placebo treated patients with the size of the difference between the placebo and antihistamine patients fairly consistent across the five hours of measurements.

18.4 A model for this experiment is given here:

$y_{ijk} = \mu + \alpha_i + \pi_{j(i)} + \beta_k + \alpha\beta_{ik} + \epsilon_{ijk}$, where

y_{ijk} is the percentage inhibition of the *jth* patient during hour k under the *ith* treatment.

α_i is the fixed effect of the *ith* treatment, $\sum_i \alpha_i = 0$

$\pi_{j(i)}$ is the random effect of patient j at *ith* treatment, iid $N(0, \sigma_P^2)$ r.v.'s

β_k is the fixed effect of Hour k, $\sum_i \beta_i = 0$

$\alpha\beta_{ik}$ is the fixed interaction effect of *ith* treatment with Day k, $\sum_i \alpha\beta_{ik} = 0, \sum_j \alpha\beta_{ik} = 0$

ϵ_{ijk} is the random effect of all other factors on percentage inhibition, iid $N(o, \sigma_\epsilon^2)$ r.v.'s

The AOV table is given here:

Source	DF	SS	MS	F	P
Treatment	1	3994.2	3994.2	12.95	0.0021
Patient(Treatment)	18	5553.2	308.5	*	*
Time	4	3467.0	866.7	6.00	0.0003
Treatment*Time	4	870.9	217.7	1.51	0.2083
Error	72	10394.4	144.4		
Total	99	24279.7			

There is a significant difference ($p - value < 0.0001$) in the mean percentage between the treated and placebo patients. The size of the difference is consistent across the five hours because there is not a significant interaction between treatment and hours (p-value=0.2083).

18.5 Defining onset as the first time at which a significant reduction in this reaction occurs, a Bonferroni t-test can be used to detect onset. We need to test at each hour if the mean percentage inhibition is greater than 0 using the test statistic:

$$t = \frac{\bar{y}_{i.k} - 0}{\sqrt{MSE}/\sqrt{n_{ik}}} = \frac{\bar{y}_{i.k} - 0}{\sqrt{141.12}/\sqrt{10}}.$$

Since we are making 10 tests, we will use $\alpha = \frac{.05}{10} = .005$ and critical value $t_{.005,72} = 2.646$. We will declare that the mean percentage inhibition is greater than 0 whenever $t > 2.646$. The results are summarized in the following table:

	Time				
	1	2	3	4	8
Antihistamine(Mean)	20.70	28.57	31.24	29.44	25.63
t-value	5.51	7.61	8.32	7.84	6.82
Significant	Yes	Yes	Yes	Yes	Yes
Placebo(Mean)	-0.76	12.55	18.23	24.79	17.57
t-value	-0.20	3.34	4.85	6.60	4.68
Significant	No	Yes	Yes	Yes	Yes

Onset occurs during the first hour for the antihistamine treatment whereas onset occurs at the second hour for the placebo. The antihistamine has an earlier onset and a consistently larger mean percentage inhibition than the placebo at each hour.

Supplementary Exercises

18.7 a. The mean and standard errors are given here:

Sequence	Period	Mean	Standard Error of Mean
1	1	7.3250	0.3594
1	2	6.850	0.2667
2	1	7.150	0.1955
2	2	7.5375	0.2427

b. A plot of the sleep duration data is given here:

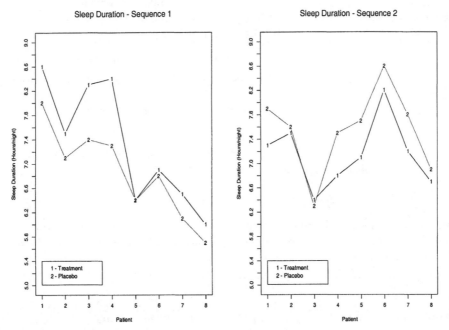

c. The AOV table is given here:

Source	DF	SS	MS	F	P
Sequence	1	0.5253	0.5253	0.46	0.5087
Patient(Sequence)	14	15.8619	1.1330	*	*
Treatment	1	1.4878	1.4878	26.29	0.0002
Period	1	0.0153	0.0153	0.27	0.6114
Error	14	0.7919	0.0566		
Total	31	18.6822			

There is a significant difference ($p-value < 0.0001$) in the mean sleep duration between the treated and placebo patients. There is not a significant Sequence nor Period effect.

18.11   a. A plot of the AUC means is given here:

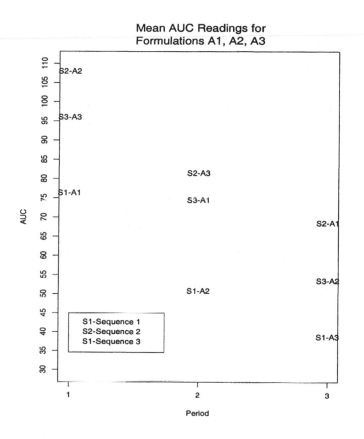

Mean AUC Readings for
Formulations A1, A2, A3

b. Yes, for each of the three formulations, the mean AUC is higher in Period 1 in comparison to the other two periods.

c. No, the relative ordering of the three formulations is different in each of the three periods. In Period 1, A2 has the largest mean AUC, in Period 2, A3 has the largest mean AUC, and in Period 3, A1 has the largest AUC mean.

18.12 The analysis of variance table is given here:

Source	DF	Adj SS	Adj MS	F	P
Sequence	2	7246.8	3623.4	3.33	0.071
Patient(Sequence)	12	13072.2	1089.4	*	*
Formulation	2	41.7	20.8	0.31	0.733
Period	2	12110.3	6055.1	91.21	0.000
Error	26	1726.0	66.4		
Total	44	34197.0			

Based on the results in the AOV table, the conclusions based on the profile plot are confirmed. There is a significant Period effect ($p - value < 0.0001$), the effect due to Formulations is not significant, ($p - value = 0.733$), and there is not an effect due to Sequence ($p - value = 0.071$).

191

18.13 The analysis of variance table for the Period 1 data is given here:

```
Source        DF    Adj SS    Adj MS      F      P
Formulation    2    2553.8    1276.9    2.51   0.123
Error         12    6112.6     509.4
Total         14    8666.4
```

A similar result is obtained using just the first Period, there is not an effect due to due to the formulations ($p - value = 0.123$). The crossover design used in Exercise 18.12 is more suitable because the variability in response from individual patients is reduced by having each patient respond to each of the three formulations.

18.17   a. The mean efficiencies for each model at each time point are given here:

Model	Time				
	1	2	3	4	5
1	1.72	1.68	1.75	1.76	1.72
2	1.95	1.90	2.01	2.05	1.99

A plot of the mean efficiences is given here:

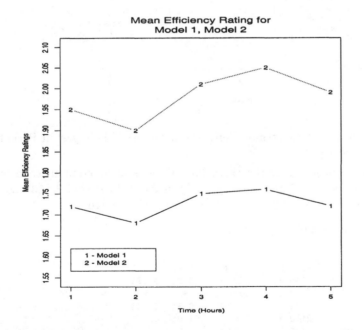

b. The analysis of variance table in the textbook yields the following results:

The adjusted p-value for the Time by Model interaction is 0.7528. Thus, there is not significant evidence of an interaction.

The p-value for the main effect of Model is 0.0960 which indicates that there is not significant evidence of a difference between the two models with respect to mean efficiency ratings.

The adjusted p-value for the main effect due to Time is 0.0512 which indicates there is not significant evidence of a difference in the mean efficiency ratings across the five time periods.

c. The correction factors increase the p-values for the Time factor. Using the uncorrected p-values, there is a significant effect due to Time which is contradicted by the adjusted p-values.

Chapter 19: Analysis of Variance for Some Unbalanced Designs

19.2 A Randomized Block Design with One or More Missing Observations

19.1   a. Complete Model: $y_{ij} = \mu + \alpha_i + \beta_j + \epsilon_{ij}$; $i = 1, \cdots, 4$; $j = 1, \ldots, 5$.

Reduced Model: $y_{ij} = \mu + \beta_j + \epsilon_{ij}$; $i = 1, \cdots, 4$; $j = 1, \ldots, 5$.

  b. The AOV table for testing treatment effects is given here:

Source	df	SS	MS	F	p-value
Treatment(adj)	3	140.8008	46.934	904.414	.0001
Block	4	2.6147			
Error	11	0.5708	0.051894		

The $p-value < 0.0001$ for treatment indicates that the treatment means are significantly diffferent.

19.2 $LSD = t_{.025,11}\sqrt{2MSE/b} = (2.201)\sqrt{2(.051894)/5} = 0.317$ for comparing treatments without any missing observations.

$$LSD = t_{.025,11}\sqrt{MSE\left(\frac{2}{b} + \frac{t}{b(b-1)(t-1)}\right)} = (2.201)\sqrt{(.051894)\left(\frac{2}{5} + \frac{4}{5(5-1)(4-1)}\right)} = 0.343$$

for comparing treatments with a missing observations.

The treatment sample means and comparisons are given here:

Treatment	1	2	3	4*
Means	15.26	9.46	9.36	8.377
Groupings	a	b	b	c

* Treatment 4 is missing an observation

The only pair of treatments which are not significantly diffferent are treatments 2 and 3.

19.5 From fitting the model $y_{ij} = \mu + \alpha_i + \epsilon_{ij}$, obtain $SSE_2 = 141.37167$.

The increase in SSE due to dropping the block effect from the model is then computed as

$SSB_{adj} = SSE_2 - SSEcomplete = 2.68350 - 0.57083 = 2.11267$,

which is the value listed as the DIARY Type III SS in the AOV for the complete model.

19.7   a. The AOV's for the complete and reduced models applied to the data set without estimating the missing values are given here:

```
Complete Model: Response versus Mixture, Investigator

Source     DF    Adj SS    Adj MS     F       P
Mixture     3    210236    70079   1014.10  0.000
Investig    4       345       86      1.25  0.347
Error      11       760       69
Total      18    219907
```

```
Reduced Model: Response versus Investigator

Analysis of Variance for Response, using Adjusted SS for Tests

Source     DF    Adj SS    Adj MS     F      P
Investig    4      8910      2228    0.15  0.961
Error      14    210997     15071
Total      18    219907
```

$SST_{adj} = SSE_{reduced} - SSE_{complete} = 210997 - 760 = 210237$, with df=14-11=3.

$SSB_{adj} = TSS - SST_{adj} - SSE_{complete} = 219907 - 210237 - 760 = 8910$, with df=18-3-11=4.

Summarize these values in an AOV table:

Source	df	SS	MS	F	p-value
Treatment(corrected)	3	210237	70079	1014.3	.0001
Block	4	8910			
Error	11	760	69.1		
Total	18	219907			

The above AOV table yields identical conclusions as the conclusions reached in Exercise 19.6.

b. Use the method outlined above, that is, fit complete and reduced models to the observed data without estimating the missing values.

19.3 A Latin Square Design with Missing Data

19.8 $SSE = 1.44$ with df=11

$SST_{adj} = SSE_{reduced} - SSE_{complete} = 166.94 - 1.44 = 165.50$, with df=15-11=4.

19.9 a. To test for a treatment effect, $F = \frac{165.50/4}{1.44/11} = 316.1$, with

$p - value = Pr(F_{4,11} \geq 316.1) < 0.0001$

There is significant evidence of a difference in mean elongation of the four versions.

b. For comparing pairs of treatments not having missing observations:

$$LSD = t_{\alpha/2, df_E} \sqrt{\frac{2MSE}{t}} = 2.201\sqrt{(2)(0.131)/5} = 0.503$$

For comparing pairs of treatments having missing observations:

$$LSD = t_{\alpha/2, df_E} \sqrt{MSE(\frac{2}{t} + \frac{1}{(t-1)(t-2)})} = 2.201\sqrt{(0.131)(\frac{2}{5} + \frac{1}{(5-1)(5-2)})} = 0.554$$

The treatment sample means and comparisons are given here:

Version	D	A	B	C	E*
Means	17.72	19.64	22.98	23.88	24.77
Groupings	a	b	c	d	e

* Version E is missing an observation

All pairs of versions have significantly diffferent mean elongations.

Supplementary Exercises

19.12 a. Balanced incomplete block design with 3 of 6 treatments appearing in each of the 10 blocks.

b. $t = 6, \quad b = 10, \quad r = 5, \quad k = 3, \quad n = 30, \quad \lambda = 2$

19.13 The following table contains the intermediate calculations needed to obtain the sum of squares for the treatment:

Block	Block Total	Block Mean
1	106	35.33
2	125	41.667
3	115	38.333
4	115	38.333
5	107	35.667
6	157	52.333
7	142	47.333
8	116	38.667
9	154	51.333
10	127	42.333

Treatment	A	B	C	D	E	F	Total
$y_{i.}$	211	175	284	172	171	251	1264
B_i	642	580	695	595	640	640	
$3y_{i.} - B_i$	-9	-55	157	-79	-127	113	0
$(3y_{i.} - B_i)^2$	81	3025	24649	6241	16129	12769	62894

$\bar{y}_{..} = 1264/30 = 42.133$

$$TSS = \sum_{ij}(y_{ij} - 42.133)^2 = 3235.467$$

$$SSB = k\sum_j(\bar{y}_{.j} - \bar{y}_{..})^2 = 3\sum_j(\bar{y}_{.j} - 42.133)^2 = 1034.80$$

$$SST_{adj} = \frac{t-1}{nk(k-1)}\sum_i(ky_{i.} - B_{(i)})^2 = \frac{6-1}{(30)(3)(3-1)}(62894) = 1747.056$$

$$SSE = TSS - SST_{adj} - SSB = 3235.467 - 1747.056 - 1034.8 = 453.611$$

Summarizing in an AOV table:

Source	df	SS	MS	F	p-value
Treatment(ADJ.)	5	1747.056	349.411	11.55	.0001
Block	9	1034.8			
Error	15	453.6111	30.241		
Total	29	3235.467			

Because the $p-value < 0.0001$, we conclude that there is significant evidence that the six antihistamines have different mean responses.

19.14 The adjusted treatment means are obtained from the equation:

$$\hat{\mu}_i = \bar{y}_{..} + \frac{ky_{i.} - B_{(i)}}{t\lambda} = 42.133 + \frac{3y_{i.} - B_{(i)}}{(6)(2)}$$

$$MSE = 30.241 \quad df_{Error} = 15 \quad t_{.025,15} = 2.131$$

$$LSD = t_{.025,15}\sqrt{\frac{2kMSE}{t\lambda}} = 8.29$$

The calculations are summarized in the following table:

Treatment	A	B	C	D	E	F
$\bar{y}_{i.}$	42.2	35	56.8	34.4	34.2	50.2
$3y_{i.} - B_i$	-9	-55	157	-79	-127	113
$\hat{\mu}_i$	41.38	37.55	55.22	35.55	31.55	51.54

The groupings based on LSD are given here:

Treatment	E	D	B	A	F	C
$\hat{\mu}_i$	31.55	35.55	37.55	41.38	51.54	55.22
Groups	a	ab	ab	b	c	c

The treatments with common letter are not significantly different. Thus, the significantly different pairs of treatments are:

(E,A), (E,F), (E,C), (D,F), (D,C), (B,F), (B,C), (A,F), (A,C),

19.15 Minitab output is given here:

```
General Linear Model: Response versus Person, Treatments

Source     DF    SS         MS       F      p-value
Person     9     1034.80    56.98    1.88   0.134
Treatmen   5     1747.06    349.41   11.55  0.000
Error      15    453.61     30.24
Total      29    3235.47
```

Least Squares Means for Response

Treatmen	Mean	SE Mean
A	41.38	2.703
B	37.55	2.703
C	55.22	2.703
D	35.55	2.703
E	31.55	2.703
F	51.55	2.703

The sum of squares and estimated treatment means agree with the calculations given previously.

19.16 Minitab output is given here:

General Linear Model: Rating versus Consumer, Pilow

Source	DF	Seq SS	Adj MS	F	P
Consumer	11	4575.00	41.30	1.18	0.369
Pilow	8	11727.33	1465.92	41.98	0.000
Error	16	558.67	34.92		
Total	35	16861.00			

Least Squares Means for Rating

Pilow	Mean	SE Mean
A	61.50	3.364
B	20.17	3.364
C	40.83	3.364
D	75.83	3.364
E	86.83	3.364
F	70.83	3.364
G	75.50	3.364
H	50.17	3.364
I	35.83	3.364

The results from Minitab agree with the calculations given in Example 9.5.

19.17 The method of complete and reduced models could be used to evaluate the block effect. First fit the complete model:

$y_{ij} = \mu + \alpha_i + \beta_j + \epsilon_{ij}$ and obtain $SSE_{complete}$ and $df_{complete}$.

Next, fit the reduced model (remove block effect):

$y_{ij} = \mu + \alpha_i + \epsilon_{ij}$ and obtain $SSE_{reduced}$ and $df_{reduced}$.

$SSB_{adj} = SSE_{reduced} - SSE_{complete}$ with $df_{block} = df_{reduced} - df_{complete}$.

The test for a block effect is $F = \frac{SSB_{adj}/df_{block}}{MSE_{complete}}$ with $df = df_{block}, df_{E,complete}$

Reject H_o and declare there is significant evidence of a block effect if $F \geq F_\alpha$

19.21 a. The number of defects, y_{ijk}, observed by Inspector k on Assembly Line j having Training level i.

198

b. This is a repeated measures experiment with repeated observations by Inspector, treatment factor: Level of Training and replication source: Assembly Line. A model for this experiment is

$$y_{ijk} = \mu + \alpha_i + \pi_{j(i)} + \beta_k + \alpha\beta_{ik} + \beta\pi_{j(i)k} + \epsilon_{ijk}, \text{ where}$$

μ is overall mean number of defects

α_i is the fixed effect of ith level of training

$\pi_{j(i)}$ is the random effect of the jth assembly line at the ith training level

β_k is the random effect of the kth inspector

$\alpha\beta_{ik}$ is the random interaction effect between the ith level of training and the kth inspector

$\beta\pi_{j(i)k}$ is the random effect of the interaction between the jth assembly line at training level i with the kth inspector

ϵ_{ijk} is the random effect of all other factors

c. An AOV table with df and expected MS is given here:

Source	df	Expected MS
Training	1	$\sigma_\epsilon^2 + 2\sigma_{L(T)*I}^2 + 6\sigma_{T*I}^2 + 12\sigma_I^2 + 4\sigma_{L(T)}^2 + 12Q_T$
Line(Training)	4	$\sigma_\epsilon^2 + 2\sigma_{L(T)*I}^2 + 4\sigma_{L(T)}^2$
Inspector	1	$\sigma_\epsilon^2 + 2\sigma_{L(T)*I}^2 + 12\sigma_I^2$
Training*Inspector	1	$\sigma_\epsilon^2 + 2\sigma_{L(T)*I}^2 + 6\sigma_{T*I}^2$
Line(Training)*Inspector	4	$\sigma_\epsilon^2 + 2\sigma_{L(T)*I}^2$
Error	12	σ_ϵ^2
Total	23	

19.22   a. We can use a mixed models approach to test the relevant hypotheses.

b. The interaction between Training and Inspector and the main effects due to Training and Inspector. Since the random effect of Line(Training)*Inspector is essentially neglible, the term $\sigma_{L(T)*I}^2$ will be set to zero. In fact, $\hat{\sigma}_{L(T)*I}^2 = \frac{MS_{L(T)*I} - MSE}{2} = \frac{1.333 - 105}{2} = -51.83$, which is set equal to 0. Taking into account the resulting change in the expected MS's, we obtain the following test statistics:

Training*Inspector: $F = \frac{MS_{T*I}}{MSE} = \frac{1.5/1}{105/12} = 0.17 \Rightarrow p - value = 0.6861 \Rightarrow$

There is not significant evidence of an interaction effect between Inspectors and Training.

Training: To determine the test statistic for testing the main effect due to training we need to examine the expected MS column. We note that under the null hypothesis of no main effect due to Training, $Q_T = 0$. This implies that under the null hypothesis of no main effect due to training that

$EMS_T = EMS_{L(T)} + EMS_I + EMS_{T*I} - 3 * EMSE$

Thus, the denominator of our test statistic is

$MS_{L(T)} + MS_I + MS_{T*I} - 3 * MSE = 14.17 + 0.67 + 1.5 - 3(8.750) = -9.91$.

This would be meaningless, hence we will use as an approximation the following test statistic.

$F = \frac{MS_T}{MSE} = \frac{130.67}{8.750} = 14.93 \Rightarrow p - value = 0.0023$.

There is significant evidence of an effect due to Training. That is, the Additional Training appears to have reduced the mean number of defects.

Similarly we obtain there is not a significant effect due to Inspectors ($p-value = 0.7872$).

c. A profile plot of the mean number of defects for the levels of training is give here:

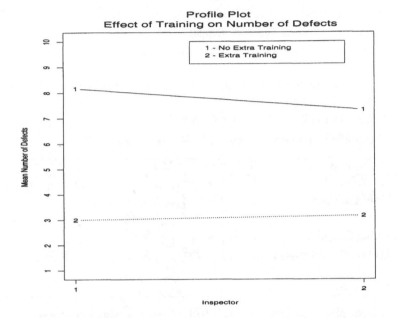

19.23　a. No, the model does not change. We can analyze the data by fitting the complete model and then a reduced model in which we remove the term for Training. Then use the change in SSE between the two models to determine the significance of Training.

b. Since the main effect of Training becomes embedded in all the interaction and nesting terms involving Training, it is not possible to test for this term using the results given in the output provided in the textbook. This can be observed by noting in the output that the df for Model are the same for all three models. Also, the Sum of Squares for Error are the same for all three models. Thus, if we attempted to compute SSE_C we obtain 0. This would contradict our intuition. There was a very significant difference in the main effect due to training in the analysis of the complete data set and dropping two data values should not completely alter this result. The correct analysis can be obtained by using the following computer output which uses an adjusted SS to compensate for the missing values.

General Linear Model: Response versus Inspector, Training, Line

Analysis of Variance for Response, using Adjusted SS for Tests

Source	DF	Seq SS	Adj SS	Adj MS	F	P
Inspecto	1	2.23	0.04	0.04	0.00	0.954
Training	1	125.67	124.32	124.32	12.13	0.006
Inspecto*Training	1	0.15	0.89	0.89	0.09	0.774
Line(Training)	4	50.14	48.75	12.19	1.19	0.373
Inspecto*Line(Training)	4	2.62	2.62	0.66	0.06	0.991
Error	10	102.50	102.50	10.25		
Total	21	283.32				

From the above output we can observe that the level of significance of the main effect due to Training has been reduced somewhat from a p-value = 0.0023 to a p-value = 0.006. However, the effect is still highly significant.